FINANCING OF ENERGY SECTOR IN DEVELOPING COUNTRIES

About the Editor

Pradeep Chaturvedi is a well known expert in the Energy Sector. His expertise lies in project planning, implementation and management with emphasis on the South Asia Region. A number of pioneering projects and studies of national and international significance have been conducted under his guidance. His involvement with the FAO's Regional Wood Energy Development Programme, based at Bangkok, has given him exposure to biomass use for energy in the South Asia Region.

Mr. Chaturvedi is the Vice-Chairman of the UNESCO sponsored World Renewable Energy Network based in UK. He is President of the Bio Energy Society of India, and Secretary General of the Indian Association for the Advancement of Science. He is also a member of the Rural Energy Study Committee of the World Energy Council. Mr. Chaturvedi is a graduate in Enginnering. He is a Fellow of the Institution of Engineers (India) and Fellow of the Institute of Directors. He is also the recipient of FAO's World Food Day Medal, 1991.

Mr. Chaturvedi has edited and authored 24 books related to Sustainable Development, Rural Energy and Energy Conservation, and has contributed papers at a number of national and international conferences.

FINANCING OF ENERGY SECTOR IN DEVELOPING COUNTRIES

(Proceedings of the International Conference on Financing of Energy Sector in Developing Countries, Organised by the Indian Members Committee-WEC, under The Institution of Engineers (India), during July 15-16, 1998 at New Delhi)

Editor
PRADEEP CHATURVEDI

World Energy Council
CONSEIL MONDIAL DE L'ENERGIE

Published for

THE INSTITUTION OF ENGINEERS (INDIA)
DELHI STATE CENTRE

By

CONCEPT PUBLISHING COMPANY, NEW DELHI-110059

ISBN 81-7022-747-X

First Published 1999

Published and Printed by

Ashok Kumar Mittal
Concept Publishing Company
A/15-16, Commercial Block
Mohan Garden
New Delhi-110059 (India)
Phones : 5648039, 5649024
Fax No. 091-11-5648053

CONTENTS

FOREWORD

The 16th Congress of the World Energy Council held at Tokyo in 1995 expressed deep concern at the deprivation of energy sources for 2 billion people in developing countries, 800 million of whom reside in South Asia region. Subsequently, the World Energy Council sponsored a Study on Financing of Energy Sector. This Study brought out various issues on financing of energy sector in developing countries with the focus on domestic capital markets and policy measures. Realising the importance of the issues raised in the Report, the Indian Member Committee decided to organise an Energy Forum on *"Financing of Energy Sector in Developing Countries"* with the help and cooperation of World Energy Council and Government of India, through the Ministry of Power and other energy ministries.

I have pleasure in presenting the proceedings of the Conference to the Executive Assembly at the 17th Congress of the World Energy Council at Houston.

Prof. **A.K. GHOSE**
Chairman
Indian Member Committee
World Energy Council

September 1, 1998

INTRODUCTION

An International Conference on 'Financing of Energy Sector in Developing Countries' was organised by the Indian Member Committee of the World Energy Council during July 15-16, 1998 at Hotel Taj Mahal, New Delhi. The Conference was inaugurated by Mr. P.R. Kumaramangalam the Hon'ble Minister for Power, Government of India. The inaugural session was presided over by Mr. Dilip Ray, the Hon'ble Minister for Coal, Government of India. The Conference was a major event for South Asia, preparatory to the 17th Congress of the World Energy Council, being organised at Houston, USA, during September 13-18, 1998. Over 300 delegates from 15 countries representing government ministries, public private energy enterprises, Confederation of Indian Industries, Independent Power Producers, Central WEC Office and Regional WEC Vice Chairman for South Asia and regional WEC member committees participated in the Conference. The participants welcomed the IMC initiative to dialogue directly with the energy industry to address collectively the challenges of financing energy operations in developing countries, and, in particular, the WEC member countries of the South Asia Region.

The World Energy Council in its recent Study on Financing of the Energy Sector focused on the need for mobilising domestic capital finance in each developing country to meet its financing needs. The report also focused on various policy measures that create conducive environment for rapid growth of the energy sector. Realising that the South Asia Region, which houses, about 800 million of the 2 million people without access to commercial energy, the IMC-WEC considered appropriate to focus on the financing aspect in this region.

Various discussions held in preparation for International Conference on Financing, received suggestions from the experts that the Conference should not be focussing only on hard core financing issues. It should rather deal with various policy issues that have an impact on financing of the projects. Accordingly, the Conference was organised to fill the void as to identify by the WEC so that the outcome of the Conference becomes complementary to the WEC Study on Financing of the Energy Sector.

The Conference was organised in the following Technical Divisions :

Division-I "Regional and Country Status Reports with Emphasis on Financing."

Division-II "Over-view Presentation on WEC Study on Financing of the Energy Sector and WEC Energy Facilitation Programme".

Division-III "Accessing Domestic Capital Markets"

Division-IV "Accessing International Capital Markets"

Division-V "Restructuring of State Electricity Boards and Energy Utilities."

In addition, two Panel discussions were also organised; the first one, as a starter for discussion on technical issues, and the second one to formulate long term guidelines for the follow-up action. These two Panels were as follows :

Panel-I "Impact of Fuel supply on Financing"

Panel-II "Future Strategies : Reforms, Regulations and Re-Structuring of Energy Sector to Ensure Adequate Financing".

Besides the written texts received from authors in advance, a large number of experts contributed significantly in their oral presentations and interventions. The total proceedings were tape recorded and transcribed. After editing the transcription the presentations and interventions in the form of specific sessions have been compiled to give a clear insight of the kind of discussion that took place.

We are grateful to Lt. Gen. J.S. Ahluwalia for providing necessary guidance.

We have pleasure in presenting proceedings of the International Conference on Financing of Energy Sector in Developing Countries, to the WEC Executive Assembly at Houston in September 1998 to provide information and materials that focuses on the approach in South Asian Region. The IMC-WEC is privileged to have undertaken this task of supplementing WEC's efforts in preparing for the 17th Congress of WEC.

Pradeep Chaturvedi
Editor and *Co-Chairman*
International Conference-IMC-WEC

September 11, 1998

NEW DELHI DECLARATION

Burgeoning energy demand in the developing countries call for concomitant higher investments in view of the urgent need to grow at a fast pace to fulfil socio-economic needs in the developing world. The general perception appears to be that sufficient finance would be available for the global energy sector requirements, but still there is an increasing problem in mobilising finances for energy investments in many developing countries, where 90% of the future energy demand growth is expected to occur. The instant remedy of restructuring of the economy from government control to market orientation has had its own mixed performance in different regions and different countries. The recent market crash in East Asia has alarmed nations to be more cautious in their effort to open their economies at fast pace.

The proponents of market economy have often impressed on the developing countries to create an environment conducive to attracting foreign investments. This concept has different connotations for different locations in the world. A general observation is that the developing countries needing private sector investments are unable or unwilling to make the fundamental changes in their policies and approaches as there is no guarantee that with the changes in place the private sector investments from the domestic sector and the international sector would flow in.

The Indian Member Committee (IMC) of the World Energy Council (WEC) organised an International Conference on Financing of Energy Sector in Developing Countries in New Delhi, India on 15-16 July, 1998. Over 300 delegates from 15 countries representing government ministries, public private energy enterprises, Confederation of Indian Industries, Independent Power Producers, Central WEC Office and Regional WEC Vice Chairman for South Asia and regional WEC member comittees participated in the Conference. The participants welcomed the IMC initiative to dialogue directly with the energy industry to address collectively the challenges of financing energy operations in developing countries, and, in particular, the WEC member countries of the South Asia region.

The essential requirement for mobilising financial resources in energy sector is based on consolidation and enhancement of the capability of WEC and member countries for providing technical assistance in energy planning, management, regulation and institution building. Special effort and provisions have to be directed towards the energy needs of the segment of the population which is still without access to commercial energy. It should be clearly understood that normal market forces will not achieve this and government intervention will be imperative.

The responsibility for creating an enabling business environment for good governance is now referred by international financing agencies as the responsibility of the State. The key elements that help to attract finance include sectoral policies including market pricing based on reducing subsidies over time to reflect full costs, transferring legal and regulatory framework of pricing/tariffs policies, bankability of investments, creation of effective domestic capital markets and institutional capabilities and government commitment to sustain an effective actions are essential for long term planning and change are issues to be addressed.

Perceptions differ while the situation is observed in the developing countries. Stipulations of the developed countries have an overriding effect as they specify the conditions of the developed countries.

The present status of power sector in a number of developing countries reflects inherent and integrated vicious circle of inefficiencies in the system making the problem doubly compounded, it is extremely necessary to evolve an appropriate mechanism which will balance the market approach and developing countries' system of governance.

CONCLUSION

1. Sustainable energy development is the key element for sustainability of the developing world, especially the South Asia Region where half of the global poor population live. The hard reality is that the 2 billion people in the developing world are still deprived of commercial energy supplies even at a subsidised rate.

2. Attracting financing, both domestic and external investment in energy, requires that governments are committed to take practical actions to implement appropriate policies to develop their energy sectoRs. The first requisite for attracting financing in the power sector is the development of a long term strategy. Given that in developing countries of the region most of the power is sold far below cost by public monopolies, government will need to formulate a long term strategy and a series of sequential practical actions, over a reasonable time frame to reform their energy sector by putting in place right energy policies, prices, taxes, regulatory frameworks and create enabling business environment. This means letting the energy enterprises—both public and private—to be managed as commercial enterprises by reshaping the energy industry to allow competition, wherever possible.

Commercialisation of energy enterprises and attracting financing to energy require formulation of government policies for unbundling sharing and management of risks that are applicable to all potential investors in the energy sector. Project financing involves many risks—commercial risks, market risks and governance risks. Responsibility for each must be assigned to the party best able to control and manage them—developers, lenders and/or the government.

3. Development of an effective regulatory system regarding standards and competition in the energy sector need to be developed through appropriate technical assistance. Many developing countries including India are in the process of establishing such a structure. Key regulatory reforms could include articulation and adoption of an integrated and balanced power policy, appointment of independent regulatory agencies, liberalisation of the pricing regime, price setting on actual production costs reflecting incentives for efficiency improvements and mechanisms to monitor the financial performance of utilities.

4. Many investors recognise the efforts that governments have made to attract financing in terms of policy announcements and issuing decrees on new laws and rules of the game including regulatory frameworks. Although the policy declarations may help, inevitably, it is the practice and enforcement of these rules that is of critical importance to the potential investor. Procurement issues, contract negotiatins, bureaucratic delays in getting approvals and clearances—all these impact on perception of risks in doing business and increase the most of attracting financing. If governments are to attract financing, they need to ensure successful implementation of sector reforms including effective functioning of the regulatory authorities and project execution in a coordinated manner.

5. Financing of energy sector in developing countries is closely linked to the development of domestic capital markets. They can mobilise 100% of their investment needs. The long term energy funding requirements require developing domestic capital markets to mobilise increasing share of financing. Public Pension Funds should be effectively utilised for energy sector financing. Capital markets require marketable assets. The papers of commercially viable power companies (and other utilities) have been ideal instruments and was, in fact, the foundation of US and European Capital Market development. The long term objectives of private sector participation should be to let domestic capital markets directly finance projects, sponsor autonomous and financially viable enterprises, both public and private, without recourse to government guarantees.

6. Foreign capital/investment is available at a higher cost and with strings attached. Efficient Domestic Capital Market Development can ensure secured loans at low cost.

7. A regional mechanism for sharing resources, both physical manpower and financial needs a new mechanism to be developed which should have a workable system. Regional cooperation in terms of projects that help in exploration of resources in one country and its sharing through trade with the others can be developed on the basis of mutual investment sharing and product sharing.

8. Regional local framework, power purchase agreement and investments can be developed that can be legally and administratively enforced.

RECOMMENDATION

Regional Cooperation

1. Regional Cooperation among developing countries must be supported in terms of technical cooperation, sharing of resources—both physical and financial, and developing energy safety network.

2. Action needs to be taken for the South Asia Region (with maximum number of poor) to reach the poor and to provide them with quality power through appropriate development of energy sector at appropriate costs.

World Class Energy Systems

3. The time has come for public recognition of the problems to set up near world class efficient power systems, pricing and current shelf of best practicable technologies, time has come for a public recognition of problems, setting some goals and demanding accountability. Changes in the mindsets of technocrats can only take place through massive scale of interaction through inter-regional workshops and fora, where generic solutions can be sought besides site specific ones. Regional/Inter-regional meetings should be organised through WEC Member Country Committees.

Institutional Mechanism

4. Institutional changes are required to formulate appropriate policies, to effectively implement them and to take steps to coordinate actions both at the national and regional level. This often lack coordination amongst different agencies responsible for energy.

As a first, the WEC Member Committees should initiate action to become a pressure group in respective countries by involving energy industries, both public and private, investment banks, energy utilities and government organisations. WEC Member Committees should reinforce their efforts at national/regional levels and also support such activities in countries who are not Members of WEC so far. In fact, regional forum will become an umbrella organisation for all developing countries seek to address local and regional problems.

WORKING GROUPS

To keep up the momentum generated by the conference to facilitate actions and to get things done, the following three WEC Working Groups for Action at the national/regional level need to be set up :

(i) Working Group on Development of Domestic Capital Markets.

(ii) Working Group on Best Practices for Production/Generation and Distribution of energy and energy services.

(iii) Working Group as Commercial issues.

The members drawing on the lessons of best practices in their respective country/ region would prepare an action plan at the national/regional level to get things done to promote bankable investments in energy to attract funding.

Regional Funding Mechanism

A Regional Fund should be set up for South Asia to pool resources from multilateral, bilateral and private donor agencies to support bankable energy projects and to take action to mobilise investments for the establishment of energy projects. For effective implementation, transparency in working will be essential.

Regional Centre

A Regional Centre for South Asian Region needs to be immediately set up under the aegis of the Indian Member Committee of the World Energy Council at New Delhi and with support of governments in the regional and the committee for Developing Countries of WEC, the WEC Energy Facilitation Programme and Member Countries of WEC in South Asia region. The activities will be overseen by a Governing Council having representations from WEC Member Country Committees of the South Asia Region.

PART-I
INAUGURAL SESSION

WELCOME ADDRESS

R.N. Srivastava,
Chairman, CEA, New Delhi

I have great pleasure in welcoming the Honourable Minister for Power, Honourable Minister for Coal and distinguished participants from India and abroad to this important international Conference on "Financing of the Energy Sector in Developing Countries." Presence of Mr. Ali Attiga, Vice-Chairman, WEC-Executive Assembly responsible for the Middle-East and South East Asia Region; Chairman on the Committee of Developing countries of the WEC and the Manager, Energy Felicitation programme of the WEC is a testimony to the support extended to this event by the World Energy Council. Sponsorship of the Ministry of Power Conference is indicative of the importance attached to the event by the Government of India. Co-sponsorship from various related ministries and public sector and private sector industries has really made it a national event. Presence of participants from Latin America, Africa, Europe, Middle East and South Asia will certainly provide a global overview of the present situation. Inputs from different continents will also result in identifying the success in different continents. Participants from other developing world will have an opportunity to learn from mutual experience and also share their concerns. Over the last 50 years there has been tremendous growth in the Indian Power Sector, both in terms of its size as well as the technological adoption. However, over the years it has also revealed some serious constraints in achieving sustainable long-term energy development programme. To meet the fast changing future challenges and in order to develop future action plans for developing world the conference in timely. As Chairman of the organising committee, I appreciate the efforts of various organisations in providing necessary inputs for this conference. I am sure that deliberations and outcome of this event would be made available to interested participants in different parts of the country, and also abroad. I, once again, welcome you all to this important conference and wish you a comfortable stay and effective deliberations. No doubt the programme is very intense but we will like to provide you all comforts during your stay.

WELCOME ADDRESS

A.K. Ghose,
Chairman, IMC-WEC, New Delhi

It is my pleasure and privilege on behalf of the Institution of Engineers (India) in adding my welcome to that of Chairman, Organising Committee. As the Indian Member Committee of WEC we are proud to be associated with the WEC right from its inception. In this year of Platinum Jubilee of the WEC, we can look back with pride over the significant achievements of WEC in the energy sector globally. WEC is a unique multi-nation non-governmental organisation. Of late, WEC has been focusing its efforts in the organisation of regional workshops to look at a little microscopic level, regional and site specific problems. India was privileged to host 12th World Energy Congress in 1983 at Delhi. Recently, in 1994 and 1996, we held two very successful regional workshops, where the problems of South Asia have been focused. I would like to express my gratitude on behalf of the Institution of Engineers (India) to the Minister for Power, for the support of his Ministry in planning and implementing the conference; and to the Minister of Coal for the ready help of this Ministry. Their presence today makes the deliberations valuable. I am glad that the Indian Member Committee has been operating under the umbrella and sponsorship of the Ministry of Power. Various other Energy Ministries, public sector undertakings and private sector industries have extended all help and support in organising this conference. The Secretary, Ministry of Power, as Chairman of the National Advisory Committee has guided the national endeavour in making this Conference a success. I would like to place on record our appreciation for the vision of Dr. E.A.S. Sarma, formerly, Secretary, Ministry of Power. Presently the Secretary expenditure and his successor, Mr. V.K. Pandit, the present Secretary, Ministry of Power. I would also like to express the gratitude of the Indian Member Committee, to Mr. R. N. Srivastav, Chairman, Central Electricity Authority, for chairing the Organising Committee, and to Mr. K.N. Sinha, Member Planning of the Central Electricity.

ABOUT THE CONFERENCE

Shri V.K. Pandit, IAS
Secretary, Ministry of Power, Government of India

It is my pleasure to welcome Your Excellencies, the Union Minister of Power and the Union Minister of Coal, and distinguished delegates to this International Conference on Financing of the Energy Sector in Developing Countries. This conference is a bridge between the 16th WEC Congress in Tokyo in 1995 and the forthcoming 17th WEC at Houston in September this year. Concerns had been expressed at the Tokyo Congress on the adequacy and the availability of finance for the energy sector. The WEC had initiated a study on financing the global energy sector, The study addressed itself to the financial concerns and suggested ways to overcome them.

The highest growth in the energy sector is to take place in the developing countries. It is clear that the financing needs of this sector will not be met from government funding alone. Any postponement in implementing the energy development plans in developing countries will only enhance the prevalent inequalities in the availability and consumption of commercial energy. For us, in the developing countries, the most important issue is not technology, or energy resources, but the availability of timely and affordable finance for the various projects that have been finalised or are in the pipeline.

The theme of the 17th Congress later this year is "Energy and Technology: Sustaining World Development into the next millennium."

This development can only be sustained if energy projects and services are developed, without any further loss of time, in the developing countries. The deliberations and the formulations of this Conference are expected to be a useful input for the 17th Congress. The focus of the discussions will be the perspective of the developing countries and the constraints that they are experiencing in financing their energy development programmes. Besides identifying the investment needs and the available resources, the Conference would draw out strategies for improving domestic capital markets.

The Conference has been structured so as to cover Regional and Country Status Reports with emphasis on financing, along with the Overview Presentations on the part of WEC, perspectives of economists as well as of industry. The discussions would thereafter take up specific areas, such as impact of fuel supply on financing, accessing domestic capital market, accessing international financial markets, restructuring of State Electricity Boards and energy utilities, and future strategies for reforms and restructuring of energy sector to ensure adequate financing. The range of participants in the various discussions is very diverse covering the Government, financial institutions and the private sector. We expect a highly professional exchange of ideas and experiences. The participants have a 'hands on' experience in tackling some of the most sensitive and stubborn Issue in the energy sector. There is a healthy mix of policy-makers, financiers, producers and supplieRs.

The Ministry of Power has tried to facilitate the convergency of diverse talents and resources in making the technical sessions of this Conference meaningful. We are aware that there are no easy solutions to some of the problems that are to be discussed in this Conference. We are not only optimistic but are convinced that the solutions to these problems can only come through professional discussions and debate.

We sincerely hope that the conclusions of this conference would provide useful inputs for the developing countries and help implement their energy development plans. We are grateful to the Union Minister of Power and the Union Minister of Coal for having spared their valuable time in the midst of Parliament, to be with us and guide the deliberations.

KEY NOTE ADDRESS

Ali A. Attiga
Vice Chairman, WEC-EA,
Amman (Jordan)

I am delighted to be back in India after 11 years since my last visit. I recall vividly my first visit in October 1964, when my wife and I had an extensive tour of your great country. We were highly impressed by what we heard and witnessed during that visit. Your rich traditions your diverse population and ancient culture and your democratic institutions were, and still are, subjects of admiration and national pride. My subsequent visits helped me to better appreciate the economic and scientific achievements of modern India. It is impressive how your country managed to overcome its chronic food deficit and even achieve some surplus, despite more than 100 percent increase in its population. It is equally impressive how the industrial and service sectors have developed and diversified from their modest beginning in the 1960s to their present level of achievements. At the time of my first visit, the domestic production of typewriters was the pride of India. Today, you are designing and manufacturing practically all your hardware and software needs for domestic and export markets.

All my previous visits to India were in connection with promoting Indo-Libyan or Indo-Arab cooperation. This time I come in my capacity as World Energy Council Vice Chairman for West and South Asia. I wish to thank the National Member Committee of India for inviting me to address you in this most timely Conference on the financing of the energy sector in developing countries. It is both pleasure and a privilege to be with you today, to see old friends and to meet the younger generation of professionals in the energy and finance sectors of India and other countries in the region. I want to learn from you for my own benefit but also in order to better serve you at the forthcoming Energy Congress in Houston. But before all that I wish to make the following comments regarding the theme of this conference.

1. First of all, thanks to the rapid increase in oil prices during the 1970s: most if not all developing countries today are fully aware of the vital importance of reliable access to commercial energy for their sustainable economic and social development. They are also appreciative of their needs for national and regional energy policies designed to modernize and develop the physical and institutional infrastructure of their energy

sectors. Prior to the so called oil crisis most developing countries were totally depending on multinational oil companies for their daily needs of commercial energy. Today, they have acquired expertise and institutional capacity for managing their energy requirements.

2. Another notable benefit resulting from the oil price increase was the massive increase in oil and gas resources and the substantive reduction in exploration and production costs. In today's abundance of oil and gas in international markets people tend to quickly forget the time when oil reserves were being depleted at a rapid rate prior to the sharp increase in oil prices.

3. On the demand side, the rise in oil prices induced better demand management and greater efficiency of energy utilization to the point where industrial countries began to claim total decoupling of economic growth and energy consumption. Although this trend is no longer stressed, the fact remains that enormous energy resources have been saved through improved conservation measures and greater energy efficiency. Although nearly all developing countries still have a long way to go in this highly profitable direction, there is an increasing awareness of the need for more research and development in this regard. All efforts should be made to translate this awareness into effective and operational programmes at national and regional levels. Such action is particularly needed in developing countries.

4. A fourth significant benefit spurred by the rise in oil prices and the transfer of oil ownership to the exporting countries was substantial increase of trade and financial flows between oil exporting and oil importing developing countries. Prior to that, there had been little or no direct contacts or relations among developing countries concerning energy, trade and development.

5. These, ladies and gentlemen, are but some of the benefits which came out of the oil crisis of the 1970. Although we now tend to take them for granted, this was not the case at the time of the short-lived crises, when such potential gains were either ignored or denied. The few of us from oil exporting countries who were pointing to these potential benefits, were not having much success with our audience, who often tended to attribute all the economic and social ills of the world to the increase in oil prices.

6. Now, that the argument regarding the oil crisis is over, with the abundance of oil and gas everywhere, and the price of oil is less than its level, in real terms, prior to the crisis, where do we go from here? What sort of energy policy is appropriate for the needs of different developing countries? Fortunately, there is a steady increase in the number and variety of studies and policy-oriented recommendations aimed at reaching answers to this big question. What needs to be done now is more focus and

concentration on the specific needs of individual countries and their regional potential for cooperation and interdependence in the energy and financial sectors. I feel hopeful that our deliberation in this conference will make a significant contribution in this regard. Mobilizing sufficient financial resources to achieve nationwide accessibility to commercial energy is the core issue facing most developing countries.

Nearly 40 percent of the world's population are without access to commercial energy and about 90 percent of expected global energy demand come from developing countries, particularly in Asia and Africa. That is why our conference should focus on the collective challenge facing these countries. It should also emphasize the need for appropriate energy policies and programmes at national and regional levels. Here, I stress the potential for collective action, because there exist within West and South Asia abundant energy resources and adequate financial savings. What is required is to create the type of conditions conducive to regional cooperation and interdependence. West Asia will continue to be net energy exporter for a long time, while South Asia is a growing market for oil and gas. Building mutual confidence through frequent contacts and joint ventures capable of generating inter-state trade and investments will ultimately help in the peaceful resolution of outstanding differences and conflicts of national or regional origin. This in turn would enhance the vital process of regional interdependence within and between West and South Asia. Such a happy turn of events would also make the whole region an attractive place for international investment and trade. But for this to happen the region would need high-quality leadership in public affairs and in business. It would also need increasing political liberalization within many countries of the region and among them. India's experience in this difficult area of human development deserves careful study and evaluation. Other countries in the region are making some progress in this regard, but they still have a long way to go before they reach the level required to meet the challenge of globalization and the pressure for economic and political reforms.

As for financing energy needs in developing countries, the problem is not a shortage of funds. As the WEC study points out, the difficult problem is the mobilization of resources where they are most needed. Developing profitable investment outlets in the energy sector will not only encourage the use of domestic savings, but will also attract foreign investment from within the region and from the international markets. But, for this to happen there is an urgent need in most developing countries to modernize and develop their money and capital markets with quality investment instruments and a credible regulatory system. To do that, most developing countries will need more technical assistance in institutional capacity-building and in developing and retaining the service of national expertise in finance and the energy sector. Much help is also needed in the process of transfering the energy sector from public to private ownership and management.

However, such a process should be gradual and should be carried out on trial basis according to the prevailing conditions and needs of each country. The essential requirement is not the mere transfer of existing ownership, but rather, the development of efficient management which is market oriented and with credible competition from new comers.

It is clear to me that the energy sector in most countries in our region and in Africa needs more than finance. Equally important is the provision of technical assistance and policy advice. Until recently the World Energy Council has confined its activities to studies and policy recommendations regarding availability and quality of energy resources in relation to demand anticipation. But WEC Study on Financing the Global Energy Sector has covered several institutional aspects which require appropriate technical assistance.

Thanks to the efforts and dedication of late Ian Lindsay, the WEC has provided valuable technical assistance to member countries in Africa and Asia. The Energy Facilitation programme and the Regional Energy Forums are complementary channels for assisting developing countries through country specific activities and regional cooperation. More financial and human resources are now required to achieve sustainable progress in this direction. I recommend that WEC should be enabled to consolidate and enhance its capability for providing technical assistance in energy planning, management regulation and institution building. This is an essential requirement for mobilizing financial resources in the energy sector. To be more specific, there is priority need for technical assistance in the following areas :

1. Strengthening National Member Committees, where they exist, and helping to establish those committees in countries that have not yet realized their importance as national institutions for coordinating and promoting the energy sector.
2. The promotion of an regional and sub-regional energy cooperation and interdependence among neighbouring countries.
3. Development of an effective regulatory system regarding standards, safety and competition in the energy sector.
4. Special effort and special provision should be directed towards the energy needs of that segment of the population which is still without access to commercial energy. It should be clear to all that normal market forces will not do that.

I conclude, Ladies and Gentlemen, by expressing my special thanks to all member committees in West and South Asia for supporting my nomination in 1996, even in my absence from that meeting. It is an honour and a privilege to represent West and South Asia. I thank you all for this trust and will do my best to be of service to you.

Thank you all !

PRESIDENTIAL ADDRESS

Hon'ble Shri Dilip Ray,
Minister of State for Coal, Government of India

I am grateful for the opportunity to participate in this Conference on "Financing of Energy Sector in Developing Countries" organised under the aegis of the Institution of Engineers (India) and "Indian Member Committee of the World Energy Council"

Energy is critical for economic growth, social development and human welfare. Over four billion people, constituting more than three-fourths of World's population, live in the developing countries. Yet they use under one-third of the world's energy input, with per capita level well below that of the developed countries. Driven by rising populations, expanding economies and a quest for improved quality of life, energy use in developing countries has doubled in the last two decades and is expected to double again during the next fifteen yeaRs.

To meet the ever-growing demand of energy is developing countries, it is vital to have a country-specific long-term energy policy taking into consideration both the demand as well as supply side management of the energy sector. Factors like energy resources of the country and that of neighbouring regions, planned GDP growth and energy elasticity of economies, investment needs and mobilisation of financial resources, both domestic as well as foreign, are some of the key elements for formulating such energy policies.

In India, main source of supply of primary energy is coal, sharing 57% of the overall present requirement of commercial energy. Share of other sources are, oil 32% Gas 7%, Hydro 3% and Nuclear 1%. Coal will remain the dominant source of energy for several decades. Sustained development of all facets of the coal industry is, therefore, imperative for the overall economic development of our country. India's known coal resources, as of now, have been assessed at about 206 billion tonnes and are adequate to sustain annual production of level of over 500 million tonnes and more throughout the 21st century.

The rapid development in coal industry after nationalisation enabled phenomenal growth of coal based power plant capacity addition which has grown to about 55,000 MW at present rising from a level of about 8,000 MW in the early seventies. Power sector is the single largest consumer of coal and consumes about 70% of the coal produced and the share of coal based generating capacity of power is about 65%. The share is forecast to remain at this level during the next 10 years.

Until the process of liberalisation began in India, the energy sector was almost entirely under State control. Production, pricing and even demand was regulated through planned development processes. This was true for the petroleum, power as well as coal sectors. Large public sector undertakings and utilities characterised the energy production scene in the country. Resources for development of the sector came almost entirely from the budgets at the national and the State levels. It is only for the last few years that budgetary support has been gradually phased out, prices de-regulated and the sector allowed to finance its development plans through internal generation of resources and borrowings.

Reliance on own sources of financing has contributed to more efficient performance of the public sector undertakings and utilities in the coal sector. For example, Ninth Plan outlay of Rs. 13,700 crores for Coal India Limited would be financed entirely from internal resources and a major loan from World Band/EXIM Bank. For the entire sector the plan outlay of Rs. 22,500 crores has only a small component of budgetary support for regional exploration, science and technology, environmental mitigation, and for SCCL and NLC with the balance of over Rs. 18,000 crores coming from the industry's own resources.

While this scenario is true for large Central public sector undertakings like NTPC and the oil companies it is important to improve the financial viability of the State Electricity Boards through appropriate tariff regulations. There is need to develop an addition to thermal capacity of over 50,000 MW in the next decade and resources for this would have to come primarily from the private sector.

The first move towards privatisation in coal mines was taken with amendment of the Coal-Mine Nationalisation Act, 1973, in June, 1993 to permit end users like power plants, iron and steel and cement units to mine coal for captive consumption. Later on, coal washing has also been permitted for private sector participation. Such mining for captive consumption is allowed provided the end-user company holds a minimum of 26% equity in the coal mining company. 67 mining blocks with a total geological resources of 20,000 million tonnes have been identified till now for private sector participation and 19 of these have already been allotted. A policy framework has been formulated in identifying more coal blocks which can be offered to private mining. Steps have also been taken to proceed with detailed exploration of these blocks in a phased manner, so that lead time forming when these blocks are offered is minimised.

The Government have provided a number of fiscal incentives and concessions for rapid development of the infrastructure industries and facilities viz. power generation, petroleum and natural gas including exploration and production an refining. for attracting investment and mining in coal an lignite, certain incentives in line with advantages enjoyed by power and oil sector are under active consideration of the Government.

I believe that financing of energy sector in developing countries will not be difficult in view of globalising of the World economy, if certain pre-requisite conditions necessary for attracting of investment are fulfilled. These are political stability needed for security of investment, well laid down long term economic policies and market for the product with assurance of reasonable return on investment. These call for enabling environment of political and economic regime backed up by adequate initiatives of the Government of the day.

The Government is committed to bring about reforms in the infrastructure sectors including the energy sector. The different steps involved in this regard are setting up of effective regulating authorities amendment to existing Acts/Laws, appropriate risk allocation between stake holders and development of the domestic capital market.

I once again thank the organisers for giving me this opportunity to share some thoughts with the distinguished participants as how best energy sectors in developing countries grew to improve economies of the countries and quality of lives of their people.

INAUGURAL ADDRESS

Hon'ble Minister Shri P.R. Kumarmangalam,
Union Minister of Power

Distinguished Delegates,

It gives me great pleasure to be here with you to inaugurate the International conference on Financing of Energy Sector in Developing Countries. I congratulate the Indian Member Committee of the World Energy Council for taking the initiative of organising this Conference in India. It will provide a forum for sharing information, knowledge and experience among the participating countries on financing the energy sector. The ready availability and use of energy is the foundation for socio-economic development. The importance of this sector to the developing countries is even more evident, given the skewed energy consumption patterns all over the world.

Economists and statisticians have related energy consumption and energy intensity with parameters like measurable incomes, national wealth, literacy, life expectancy and infant mortality. Since energy consumption has come to be equated with wealth and well-being, shortages in the availability of energy and the inequalities and inefficiencies in its distribution are being increasingly resented. The fact of the matter is that the disparities in the access to energy services are distributed in an embarrassingly uneven manner. These disparities become accentuated as we disaggregate regions into individual countries and, eventually, into different social strata within the countries. The richest 20% of the world's population use 55% of final and primary energy, while the poorest 20% use only 5%. The per-capita use of final energy varies by a factor of 18 between countries in South Asia and those in North America. In actual terms, the variation is reported to be from 0.3 tonnes of oil equivalent per capita to 5.3 tonnes of oil equivalent per capita. The largest disparity is in the case of electricity. The richest 20% use three-quarters of all electricity, while the poorest 20% use less than 3%, globally. This brings out their extremely limited access to commercial energy in general, and to electricity in particular.

India has registered a growth rate of commercial energy of nearly 6% per annum during the last five decades, while the growth rate of national income in the same period has been between 3% and 0.5%. Although coal continues to be the basic primary commercial energy source, oil and natural gas are becoming increasingly important.

Electricity has been the most preferred form of energy with its consumption witnessing an annual growth of nearly 10% in the last five decades, compared to the overall energy consumption growth at the rate of around 6%. While the total energy consumption in the economy has increased from 84.5 MTOE in 1953-54 to 284.5 MTOE in 1996-97, the share of commercial energy has increased from 28.4% to 68.7% during this period. I am highlighting these figures merely to emphasize the point that the inevitable increase in commercial energy, particularly electricity consumption, directs us to work out implementable plans for the development of this sector urgently.

Recent studies of the World Energy Council indicate that the cumulative capital requirements for the energy sector from 1998 to 2020 range from US $13 trillion to $20 trillion. It had been estimated that the developing countries would be required to spend at the rate of almost $2 trillion a year. Developing countries invest 3 to 4% of GDP in the energy sector. The development of the energy sector infrastructure has so far been mostly public owned in developing countries. It is not possible for the public sector alone to invest adequately in the burgeoning demand of this sector. Capital markets have reportedly grown faster than total GDP for quite some time, and this trend is likely to continue. The present annual global energy investments are about 10% of international credit financing. As the capital markets are growing faster than GDP, the size of the capital market is not likely to be a limiting factor for energy sector finance. However, capital investment is likely to be diverted to areas where the returns are higher. The returns in the energy sector in the developing countries have not been comparing well with other investment alternatives. In order to ensure larger investments in the energy sector, the perceived risks for investors in this sector have to be reduced and the rates of return have to be reasonably attractive. These returns may not be possible in all projects in developing countries. It is here that the international development finance should increase its role in the development of energy sector infrastructure through priority and concessional lending. These institutions need to show a greater understanding of the social and equity issues in developing countries.

At present, India faces an energy shortage of about 8% and a peaking shortage of 11.3%. In order to meet this shortage an incremental capacity addition or over 10000 MW per year is required this would need an investment of around Rs. 45,000 crores each year. We are planning to add nearly 40,00C MW capacity in the Ninth Plan with the combined efforts of the government of India, State Governments and the private sector. We have taken a number of steps to accelerate the development of the power sector and have encouraged private investment in power generation. Through this decade, we have encouraged independent power producers. Foreign owned companies can repatriate profits without any export obligations. There is a wide choice on the use of the type of fuel.

Approval procedures have been streamlined and the number of clearances have been reduced. The Government of India delegated the power for clearance of projects upto Rs. 1000 crores (Rupees one hundred million) at the level of State Governments. Automatic approval for projects with foreign equity of less than Rs. 1500 crores has been recently introduced. While creating a conducive climate for private investment and hastening the sanctioning apparatus, we also have to identify and weed out non-performing developers to ensure that development plans do not remain in a limbo.

Our Government is conscious of the fact that one reason for the restricted growth of the power sector has been the fact that electricity prices do not reflect real costs. This had led to sickness in a large number of State Electricity Boards and impaired their ability to invest in new generation capacity. If private capital has to be attracted, we have to ensure that subsidies are restructured to reduce investment risks. Our Government has taken a bold initiative in enacting the Electricity Regulatory Commissions Act, 1998. Under this Act, the Central and State Electricity Regulatory Commissions are being set up. These Commissions will have the task of fixing an electricity tariff which will ensure a handsome return to investors in the sector. As prudent government policy, we would try to ensure that the returns from investment in the power sector are as attractive as investments elsewhere.

There is an inherent limitation on the extent to which the electricity tariff in developing countries can be raised. Tariff has a direct relationship with the ability of the consumer to pay. The poorest segments of the population have to be supported by State intervention or cross-subsidy, High tariff are also a hindrance to industrial growth and creation of employment opportunities.

While the Government in developing countries have begun to realise their onus and re-orient policies, there is also an onus on utilities in developing countries. They must improve performance and efficiency to increase investor confidence in domestic and international markets. This calls for a higher degree of professionalism in the utilities in the energy sector, better performance and better financial management. We are proud of the performance of our Central Public Utilities and have recently granted a special status to the National Thermal Power Corporation.

We welcome a much larger share of international development finance in the energy sector in India. We are already executing prestigious projects in the energy sector with the assistance of such finance. We would be anxious to tie up future projects at the earliest. The speedy development of the power sector is imperative and any delays in the availability of finance is only likely to enhance project costs. Besides delaying the availability of power. Given the shortages in electricity supply at the moment, the country is prepared to look at whatever alternative source of funding is available, to ensure the timely completion of power projects.

The issues that are to be discussed in this Conference are too well known. The problems are also well known. We are here to see what initiatives need to be taken to formulate implementable packages and also to discuss whether the energy sector in energy starved countries is a humanitarian activity which should get priority and uninterrupted assistance.

I am confident that the deliberations in this conference will go a long way in providing solutions to complex issues and also serve as a useful input for the World Energy Conference at Houston, later this year.

PART-II

REGIONAL AND COUNTRY STATUS REPORTS WITH EMPHASIS ON FINANCING

ENERGY IN LATIN AMERICA — AN OVERVIEW AND THE BRAZILIAN CHALLENGE

1

Jose Malhaes da Silva
Chairman, Commitee on Developing Countries, WEC

Once again I would like to thank the hosts of this event for giving me the opportunity to let you have my comments on the theme assigned to me — ENERGY IN LATIN AMERICA — AN OVERVIEW AND THE BRAZILIAN CHALLENGE — at such an important Forum. It is both an honour and a pleasure to be here, in your fascinating country.

What I intend to do is to give you a brief overview on the Latin American scenario and enter into a little more detail when addressing the challenges with which my country Brazil — will face in the energy field in the next few years, highlighting the electricity sector.

The region cover 38 countries that share a historical and cultural heritage since most of the area was colonised by countries in the Iberian peninsula, starting in the 15th century.

Brazil is the only country with a Portuguese background, the others are of Spanish origin with three exceptions: Suriname, which is French, Guyana English, and Guyane Dutch.

But despite this common heritage, South American countries are very different in size, economies and resources. The three largest are Brazil, Argentina and Mexico. They are responsible for approximately 75% of the area's GDP which, in 1995, was estimated at about one trillion dollars at the 1980 rate.

The population is a mixture of the original native Indians with European, Asian and African immigrants. Nature has been bountiful to the area which has abundant resources and ecosystems that represent the earth's biodiversity with a major potential of carbon dioxide absorption.

In 1995, the region's population consisted of about 475 million inhabitants.

Adding the significant informal economy throughout the area, the regional per capita income is believed to be well above the US $ 2,100 estimated in former years. However, in terms of purchasing power parities, the per capita income is close to the US $ 4,900 average world income.

I would say that the GDP tends to grow above the world average. In fact, in 1997, the World Bank figured that it will increase 4 to 5% over the 1990-2020 period. In the meantime, the population is estimated to rise from 430 to 840 million inhabitant by the year 2050.

This all very well, but our subject is energy.

Let us see how the end consumption of energy in 1995, equivalents to about 4 billion barrels of petrol, compares with figures for the rest of the world.

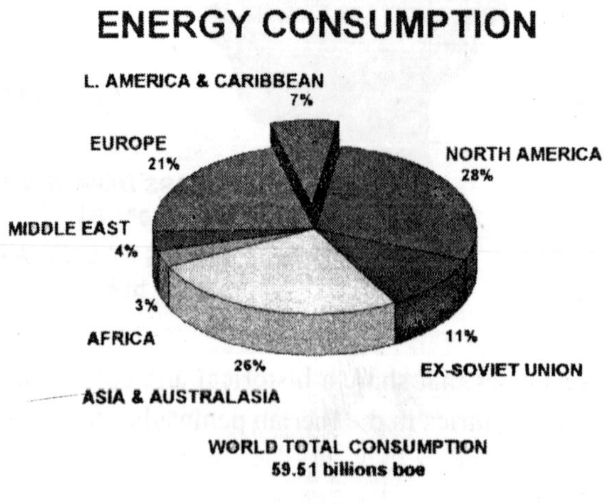

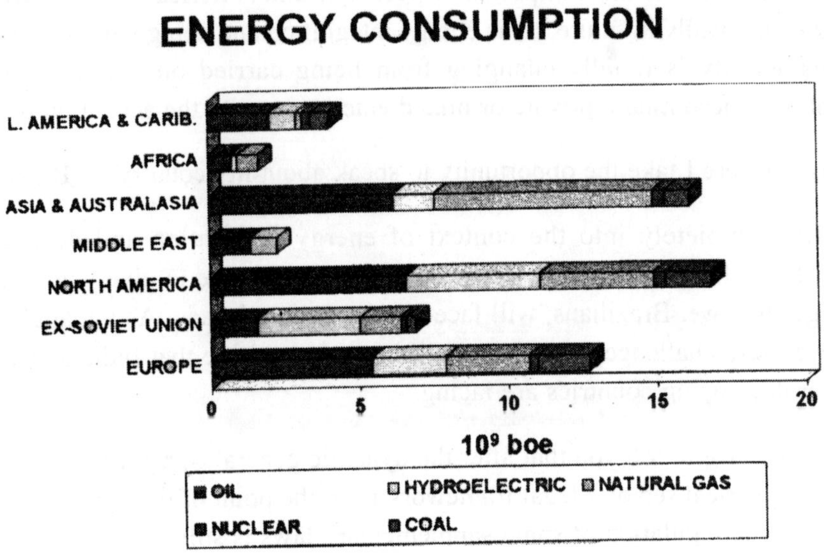

LATIN AMERICA & CARIBBEAN

World Reserves — 1995

	Latin America & Caribbean	*World*
Oil	13.0 %	991×10^9 bbl
Natural Gas	5.4 %	140×10^{12} m^3
Coal	1.6 %	1034×10^9 ton
Hydro	22.7 %	3213×10^3 MW

Note that the increase in Latin America's and the Caribbean's energy production is forecast to occur predominantly in natural gas, hydroelectricity, and renewable energies. This should have a beneficial effect by reducing air pollution in the large urban centres of the Region. Sulfur emissions are expected to drop to levels below those of 1990 within twenty years, and the forests throughout the area are potentially important as carbon sinks. It is forecast that biomass energy will be used sustainably and will not generate CO_2 emissions.

Over the period 1990 to 2020, the region will probably make investments in the range of US $ 1 trillion which is practically identical to the region's GDP in 1990 (US $ 1.1 trillion).

However, I believe the most important aspects in Latin America and the Caribbean at present and undoubtedly in future, to be energy integration and deregulation of the energy sector. This activity is rapidly changing from being carried out exclusively by the government. It is becoming a private or mixed enterprise, with the accent on private.

And this is where I take the opportunity to speak about my country — Brazil.

Brazil fits completely into the context of energy integration and deregulation I mentioned above. Let me take the electric sector as an example to give you a picture of the challenges that we, Brazilians, will face in the coming decade. As a matter of fact, I don't believe these challenges are so very different from those that India and China, as well as other developing countries are facing.

First of all, I have to tell you that after the dynamic general deregulation taking place in Brazil, the electrical sector's legal framework is on the point of being approved by our Congress: only the regulation of some supplementary laws is still pending.

The cycle of privatising distribution, generation and transmission started with the sale of distribution concessions which is now in its last stage. We are starting to put the large generating companies on the market in order to later work on the transfer of assets and transmission concessions.

At the same time, a new model is being drawn up for the electricity sector. An independent regulatory agency, ANEEL, the National Electric Power Agency, has already been implemented.

ELECTROBRAS, the holding company of federal generation and transmission companies, is still in charge of planning at national level and makes sure that the interests of the system are preserved.

Generation capacity expansion is preceded by planning studies that encompass several temporal horizons. Contemplating various possible scenarios, it is their objective to pinpoint when, where and how to increase the electric power supply in order to best meet the needs of the consumer market within criteria of guarantee, reliability, continuity and costs. To this end, marginal costs of generation expansion and the cost of an energy deficit are calculated; also, parameters pertinent to, respectively, comparative analyses of production unit costs and a policy for programming supplementary thermal generation.

Let me stress that the predominance of hydraulic generation with pluriannual regulating reservoirs belonging to different companies results from the need for integrated action with a view to electro-energetic optimisation of the system, greater efficiency, and the lowest electric power supply cost.

SOUTH AMERICA AND BRAZIL
GENERAL INFORMATION
1997

ELECTRIC POWER GENERATION
BREAKDOWN BY SOURCE
1997

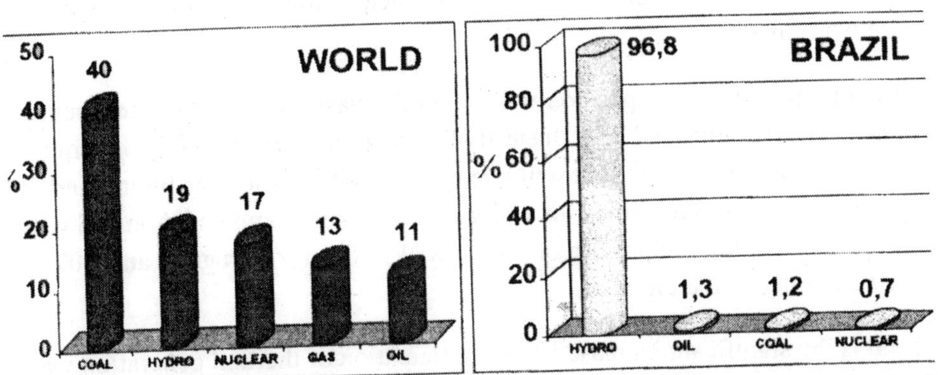

The composition of the expansion policy considers, aside from the economic-financial priority which is supply at the lowest cost possible, the following three factors that could possibly affect the start of operation of programmed projects.

- Physical-financial schedule of works underway.
- The minimum time needed for implementing new undertakings, considering engineering and bidding aspects, and based on the need for studies or designs the possible need for concession bids and the execution of the works;

- Environmental viability of the projects, both from the legal aspect of obtaining a licence, and from the social-environmental questions of each undertaking.

The following reference electric power market needs resulted from these premises:

BRAZIL 1998 - 2007

	1997	2007	ANNUAL %
CONSUMPTION (TWh)	288	480	5.2

Leading to the need for expanding the installed capacity :

BRAZIL 1998 - 2007

INSTALLED CAPACITY	*1997*	*2007*	*ANNUAL %*
HYDROELECTRIC	54.2	79.3	3.9
TERMOELECTRIC	5.1	16.4	12.4
TOTAL	59.3	95.7	4.9

Two comments that corroborate above mentioned statements on this projected expansion are worth making:

The first is that part of this expansion will be on the basis of thermal plants running on gas from Bolivia, Argentina and Peru. Instead of building new plants, Brazil is acquiring 1000 MW from Argentina, a figure that will rise to 2000 MW in the near future; and 200 MW from Venezuela. This exemplifies the energy integration I commented on and which started with the shared ITAIPU project of 14600 MW, 50% for Paraguay and 50% for Brazil, at the beginning of the eighties.

The second is the significant increase in the participation of thermal generation which requires less construction time, shorter distances for transmission lines since they are built closer to load centres, and few environmental impacts that tend to diminish even more due to technological advances. This increase of thermal generation will also endeavour to reduce the distortion caused by 97% of Brazil's electricity originating from hydraulic generation.

The following is the expected transmission and substation expansion:

BRAZIL 1998 - 2007

TRANSMISSION LINES (1000 KM)

Actual (1997)	*Future (2007)*	*% in the period*
171,3	218,1	27

BRAZIL 1998 - 2007

TRANSMISSION LINES (1000 KM)

Actual (1997)	*Future (2007)*	*% in the period*
218,8	315,4	44

If we translate the above mentioned expansion into cash, we arrive at the challenge I mentioned before.

BRAZIL 1998 - 2002

FORESEEN INVESTIMENTS - 10^9 us $

	1998	1999	2000	2001	2002	TOTAL
Generation	3.0	3.2	3.1	2.6	2.5	14.4
Transmission	3.0	2.2	1.3	1.8	1.9	10.2
Distribution	1.4	1.3	1.3	1.3	1.3	6.6
General inst.	0.4	0.4	0.3	0.3	0.3	1.7
Total	7.8	7.1	6.0	6.0	6.0	32.9

TEN-YEAR GENERATION PLAN MAIN PLANTS PROGRAMMED FOR THE FUTURE

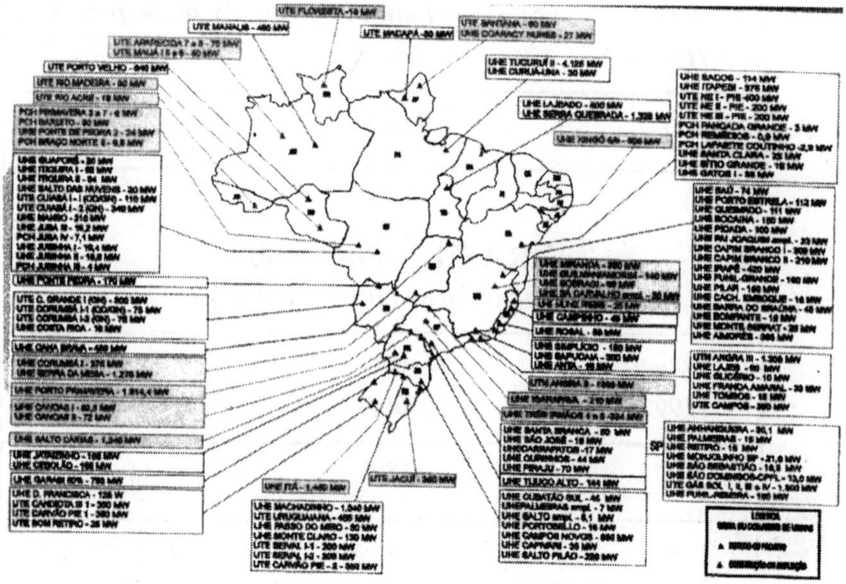

YEAR 2007 TRANSMISSION SYSTEM

This challenge will only be overcome if private enterprise is invited to participate. The need for it in order to reach the targets set is fundamental and increasing, from the

fast maturity, endeavouring to make them viable in association with the private sector and/or its own subsidiaries by means of funding, consortia, marketing energy generated in its own system and co-ordinating the implementation of larger works.

It is obvious that, in order to overcome these challenges, there is a need for a stable economy in Brazil and the world. In the context of a globalised economy, my country is doing its bit. It is, however, fundamental that strict rules be laid down for controlling international volatile capital. So far, this has not happened because it is not of interest to industrialised nations.

Once we are free of the spectres of these crises which must stopped to obtain global economic stability, I am sure that we will overcome the challenges in store for us.

Ladies and gentlemen, this is what I had to say to you.

Thank you very much !

ENVIRONMENTAL EFFECTS AND FINANCING STRATEGY OF POWER GENERATION IN ASIA TO THE YEAR 2010 AND BEYOND

Hisham Khatib
Honorary Vice Chairman, World Energy Council

PREAMBLE

Asia's electricity production is mainly dominated by that of China. After the US, China is the world's second largest electricity producer. Its electricity generation this year is expected to reach 1200 TWH which is almost 9% of global electricity production. China's production is increasing at an annual rate of 6% which is double that of Japan's and of the world average.

Main producers of electricity in Asia and their expected production in 1998 is as follows:

Table 2.1 : Demand for Electricity in Asia[1]

	TWH
China	1200
Japan	1070
India	480
Korea	210
Ukraine	200
Saudi Arabia	110
Thailand	100
Iran	100
Indonesia	70
Pakistan	60
Malaysia	55
Sum	3655
Rest of Asia (estimated)	445
	4100

Therefore Asia with an electricity production of around 4100-4200 TWH, will account for around 29% of the world's electricity production, in 1998, of around 14,400 TWH. This is impressive, however, more important is that it will continue to grow at a rate of 4%—6% Annually which is 50% higher than average world electricity growth rate.

One of the most important aspects is that demand per capita in Asia (see Annex) at 1250 KWH is only half that of world average of 2500 KWH. According to a recent EPRI study it needs a consumption of 1500 KWH per capita to rise above subsistence and making even marginal economic progress. The present modest electricity utilisation of most Asian countries shows the long way ahead of them to be able to properly develop to rise above subsistence level and make marginal economic progress. This threshold consumption of 1500 KWH per capita per annum can be considered, in the future, as the dividing line between a developing and a developed country.

ASIA'S POWER SECTOR—MAIN FEATURES

There are few main features pertaining to electricity production in Asia and its future.

Rapid Growth

This was already covered in the first few paragraphs. Electricity demand growth will continue to be high and at least equal to economic growth in purchasing power parity. This can be as high as to average 4%-6% annually in most countries of Asia (with the exception of Japan) and will continue with this average, in spite of recent economic slow down in East Asia, in order to feed presently unsatisfied demand.

Heavy Reliance on Coal

Coal will dominate electricity production in Asia. Most of This will be indigenous coal (sometimes low quality). Therefore electricity production in Asia, particularly in China and India, will continue to be mainly by coal, which is a relatively polluting and environmentally not clean, to say the least.

In 2010 China will have 430 GW of its 600 GW production facilities firing coal, making it the world's leading country in coal dependence for electricity generation, more than the US with only 322 GW. All Asian countries who are members of Asia-Pacific Economic Cooperation (APEC), including Japan, will almost treble their coal consumption over the next 18 yeaRs.

No matter what may come, Asian countries with no large oil and gas reserves, will depend on indigenous coal as the main in feed for their electricity production. Therefore, in spite of all environmental outcries, die low income countries of South Asia, particularly China and India, will gradually increase their coal dependence. The local, regional and even global implications cannot be ignored. Therefore dealing with these environmental threats is a priority for Asian countries. This can be done in many ways—

- Emphasis on efficiency of energy and electricity production and usage.
- Pollution prevention through investing in emission control technologies and introducing clean-coal-technology.
- Switching to cleaner fuels whenever available (replacing coal by gas).

Inaccessibility to electricity[3]

Many low income countries in South Asia, particularly the populous countries of India, Pakistan. Bangladesh and Indonesia, do not have an electricity supply accessible to most of their population. Two-thirds of the population of these countries are without electricity. There are, in Asia alone, more than one billion people who are not supplied with electricity in any form. Providing them with electricity is the main challenge facing these countries in their endeavour for sustainable development.

Table 2.2 : World Population Access to Electricity (1995)

Population	*Rural Population*		*Population without Electricity*	
	(million)	*(%)*	*Millions*	*%*
OECD, EE& FSU	1300	25%	—	—
Developing Countries				
China	1200	68%	60	5%
India	930	72%	650	70%
Indonesia	194	62%	128	67%
Pakistan	130	64%	78	60%
Bangladesh	120	80%	78	65%
Sub Saharan Africa	600	75%	480	80%
Other DCS	1276	60%	276	22%
World	5750	54%	1750	30%

Note : China's figures are based on officially published figures, which could be substantiated

The above figures are, in many cases, estimates. However they demonstrate the extent of inaccessibility of commercial forms of energy in general and electricity in particular, to the population of South and East Asia. This inaccessibility is only equalled in Sub-Saharan Africa.

Shortages of Investment Funds

The electricity supply industry is highly capital intensive. It is probably more capital intensive than any other sector, particularly in the developing countries. Therefore, planning and proper financial and economical evaluation of projects is important to rationalise investment and achieve economic efficiency.

Recent Studies[4] indicated that there is a need to invest $ 2280 billion (in 1993 dollars) in the electricity supply industry, world wide over the period 1995-2010, that is $152 billion on the average per annum. This is around $ 2620 billion in 1998 dollaRs. Generation projects will account for 63% of these capital investments, transmission 8.8%, and distribution 21.2%. The remaining 7% will be general expenditure, mainly concerned with control, telecommunications, and similar activities. Developing countries will account for over 58% of these investments, developed countries (OECD) 32% and the Transitional Economies (East Europe and the Former Soviet Union) for the remaining 10%.

It is estimated that over the period of 1995-2010 Asia will have to invest in 1200 GWs of new capacity at a cumulative investment costs of $1380 billion as follows (Table 2.3)

Table 2.3 : Power System Expansion and Investments in Asia (1998 dollars) 1995-2010

	Expansion GW	*Investments in 1998 dollars*
China	231	537
OECD Pacific Rim	54	192
East Asia	143	316
South Asia	122	278
West Asia	50	57
Total Asia	600	1380
World Total	1200	2620
Asia/World %	50%	52%

From the above Table, it is clear that Asia has to invest slightly more than half global investments in electric power system extension.

Countries of East Asia and South Asia alone would need to invest almost $ 600 billion over 15 years, i.e. $ 40 billion per annum. These countries, with their modest income, cannot afford much of these investments from indigenous sources. They have to undertake foreign borrowing or attract foreign investments and investors (IPPs). This necessitates restructuring of the electrical power sector of most of these countries to allow this to become possible. Without this East and South Asia will continue to suffer from power shortages that inhibit its economic and social development potential.

Usually funds for expanding the electricity supply industry comes from one or more of the following sources:

- Self-financing capability of the utility
- Government funds
- Official development assistance funding (ODA)
- Local and foreign private investments.

As explained above government funds to finance commercial infrastructure projects, like electricity power is becoming more and more scarce. Self financing capabilities of utilities in Asia are very restricted (some of them working with a deficit). Therefore we are left with ODA and private investment funding. Unfortunately ODA to developing countries, has not increased during the recent yeaRs. Therefore, the financing gap of electrical power projects is increasingly being filled by calling on foreign private capital flows through IPPs and financial markets. However foreign indirect investment, in developing countries, is increasing rapidly each year, just as portfolio equity and private debt (commercial banks, bonds, suppliers credit, etc.). This is detailed by the following Table 2.4.

Table 2.4 : Capital Flows to Developing Countries (billion nominal Dollars)

	1989	*1991*	*1993*	*1994*
Official Development Assistance (ODA)	43	62	54	55
Private Debt	12	18	45	55
Foreign Direct Investment	26	37	67	78
Portfolio	4	8	47	40
TOTAL	**85**	**125**	**213**	**228**

It is not possible to ascertain how much of the above mentioned private capital is going into power development in developing countries. However, the electricity supply industry (ESI) attracts a sizeable amount of this capital. Of the estimated recent $ 48 billion power sector financing for developing countries, only $ 19 billion came from domestic sources. Therefore, of the estimated $ 1460 billion required for ESI development in developing countries, over the next 15 years, an increasing proportion will have to be provided by international private capital. For this to be achieved, much reform (privatization and restructuring) and regulatory arrangements must take place in developing countries. Three conditions are mandatory[5]:

- Governments must be committed and guarantee a financially independent electricity supply industry.
- Electric utilities have to perform in a financially and economically viable way.
- Investors (foreign as well as local) must be convinced that they will obtain a good return on their investment, and will be able to repatriate these returns.

CONCLUSION

Although Asia presently accounts for only around 30% of the world's electricity production, however, because of its rapid growth it is going to account for half of global generation expansion. Most of this expansion will be dominated by coal firing power stations, particularly in China and India. There is dear need to develop clean-coal-technologies in these countries to allow environmentally safe handling and utilisation of this polluting fuel. Nuclear facilities are expected to figure in the generation expansion of China, Korea, Japan and Taiwan. However its role will continue to be small compared to expansion of coal facilities.

Inaccessibility to electricity and shortage of funds are the main threats to the sustainable development of the low income and heavily populated countries of South Asia. In order to overcome supply shortages and allow for needed expansion, the low income Asian countries need to attract foreign investment. For this to happen they need to introduce reforms and restructuring of their electricity supply industries in order to attract private sector investments and the independent power producers (IPPs).

Because of their limited incomes and shortages most Asian countries will put more emphasis on the expansion of the supply, rather than on dealing with its environmental threats. Therefore technological assistance to these countries, particularly the introduction of clean-coal-technologies, is badly needed. In order to attract foreign investments these countries must undertake reforms and restructuring of their electrical power industry.

NOTES

1. APEC Energy Working Group, "Regional Cooperation for Power Infrastructure, 1995.

2. "Electrification and Global Sustainability", *EPRI Journal*, Jan./Feb. 1998.

3. Khatib, Hisham, "Electricity in Developing Countries", lecture to IEE (London) on 15th April 1998.

4. Resource Dynamic Corporation for US Department of Energy :"Financing world-wide electric powers can capital markets do the job?", April 19, 1997.

5. Khatib, Hisham, "Financial and Economic Evaluation of Projects in the electricity supply industry ", a book published by the IEE (UK), 1997.

ANNEXTURE
ELECTRICITY CONSUMPTION IN ASIA
(KWH per capita—1995)

Bangladesh	57
China	637
Hong Kong, China	4850
India	339
Indonesia	263
Iran	1059
Iraq	1396
Israel	4836
Japan	6937
Jordan	1140
Korea, Dem Rep.	261
Korea, Rep.	3606
Kuwait	13185
Lebanon	1224
Malaysia	1953
Myanmar	52
Nepal	39
Pakistan	304
Philippines	525
Saudi Arabia	3906
Singapore	6018
Sri Lanka	208
Syria	698
Thailand	1199
U.A.E.	7752
Yemen	99
Asian Average	1250
World Average	**2500**

Source : World Bank Atlas (1998).

OVERVIEW PRESENTATION ON GHANA POWER SECTOR

Bright Obeng-Boampong
Manager (Finance), Volta River Authority, Ghana

INTRODUCTION

It is a pleasure that I have been nominated to make a presentation at this august assembly. Ghana power sector is under the remit of the Minister of Energy and Mines (MOME). The Minister is of Cabinet rank and is assisted by a deputy. Within the Technical Department of the Ministry, there is a Director of Power who is responsible for policy development, monitoring, coordination of the implementation of the power development programme and the National Electrification Scheme (NES). Under the Government's Vision 2020, NES seeks to provide universal access to electricity for all by the year 2020. The Technical department also promotes productive uses of electricity e.g. energy efficient bulbs.

VOLTA RIVER AUTHORITY (VRA)

The VRA was established in 1961 as a fully state-owned corporation (SOE) responsible for electricity generation and transmission. The core mandates of VRA includes the planning, development and operation of hydropower generation facilities on the Volta River and the high pressure bulk transmission of electricity from such plants for sale to needy institutions in the country particularly the mines and power utilities in the neighbouring countries. In 1987,.VRA was further mandated to distribute electricity in the northern parts of Ghana (Brong Ahafo, Northern, Upper East and Upper West Regions), a function previously undertaken by VRA's sister organisation, the Electricity Corporation of Ghana (ECG). VRA hence created a Northern Electricity Department (NED) to implement the northern distribution component of the Northern Electrification Project and became a fully vertically integrated power utility.

In furtherance of the above roles, VRA has been able to source borrowings primarily from multilateral financing agencies to install a total hydro generation capacity for public electricity supply amounting to 1072 MW (912 MW from the Akosombo Hydro Dam, 1965 and 160 MW from Kpong Hydro Dam, 1981), all plants on the Volta River. In addition to a few diesel plants, the Authority also owns and operates the 30MW Tema Diesel Plant dedicated to Ghana's industries on an emergency basis. To enable VRA optimise the yield, avoid the risk of over-drafting the Volta Reservoir and provide reliable power supply, VRA has since 1997 installed the first two 100 MW combustion turbines (200 MW) of a programmed 600 MW Combined Cycle Thermal Plant at Takoradi (The TTPP). The 100 MW Heat Recovery and Steam Generator will be commissioned by December 1998, leaving 300 MW capacity outstanding for fund mobilisation. Total capacity installed by December would be 1452 MW excluding the 140 MW gas turbine being installed by Ghana National Petroleum Corporation.

VRA also manages the national transmission grid, with the main load centre in the Accra Tema Metropolitan Area. The grid has been inter-connected with those of power utilities in neighbouring countries through a 161 KV double circuit line from Akosombo to Lome (The Communaute Electrique de Benin, of Togo and Benin) and a 225 KV line from Prestea to the Compagne Ivorien electrique (CIE) of Cote d'Invoire. Another line is also planned to interconnect the grid to Ouagadougu (SONABEL of Burkaina Faso).

The main consumers of electricity in Ghana are the VOLCO (Volta Aluminium Company) or smelter based at Tema, about 80 km from Akosombo, the Electricity Company of Ghana (ECG) which handled local retail distribution in the whole of Ghana until 1987, when it became responsible for distribution only in Southern Ghana, large mining and industrial consumers served by VRA and exports to Togo/Benin (since 1972 La Cote d'Ivoire (since 1984). NED distribution is very minimal.

ECG was created as fully-owned state electricity distribution corporation in 1963 to replace the Public Works Department's Electricity Division. In 1993, a Statutory corporations (conversion to Companies Act) was enacted to convert ECG into a public company (ECG limited) in an attempt by the government to inject further efficiency and to encourage private sector participation in the energy sector.

While the VALCO load is not anticipated to grow (about 2800 Gwh), the other markets have been growing at 10% annually over the past 10 years. The forecast is for loads to grow at between 67% at least until the year 2000. High voltage mining and industrial consumption is registering high growth in sympathy with the continuing expansion in economic production activities. Demand for electricity over the medium to long term is also growing because of the government's NES Project. Currently, 80% of the country's quantifiable energy supply is in the form of wood-fuels for domestic cooking.

An energy supply and demand imbalance exists. In 1983-84, hydro supply was constrained by severe drought, which culminated in power curtailment. The same experience has been witnessed in 1998. There is also the need to retrofit supply infrastructure and bring on stream promising mini-hydrosites. At the same time, the government intends to reduce the extent to which it is called upon to apply public investment resources to provide foreign guarantees to cover debts contracted by the public utilities for power projects. These utilities were set up to operate along commercial lines but were guided by government policy in the determination of rates. Government has accordingly introduced appropriate reforms to promote competition and investment in electricity supply via IPPs and reduce any direct government control of operations, improve efficiency and reduce government expenditure. The Statutory Corporations (conversion to Companies Act) passed in 1993 which converted ECG to a limited company and made it no more tax exempt, is one of the core elements of government strategy.

An independent Public Utilities Regulatory Commission (PURC), 1997 has ben charged with the task to approving rates that would be fair to consumeRs. An Energy Commission has also been set up in 1998 to handle licensing of entities for generation, transmission, distribution as separate businesses and to monitor industry standards.

VRA has hence registered a National Grid company for managing the open-access transmission network to facilitate direct electricity sales by IPPs to consumers, TAPCO for divestiture and operation of the TTPP plant, NEDCO to convert the Northern Electricity Department to a limited company and VOLTA TELECOM LTD. for operating the Communications facilities. Each of these subsidiary companies would require new financing arrangements.

FINANCING NEEDS

So far, much of the investments in the power sector have become possible because of borrowings from international financing organisations, with government providing adequate guarantees. With out recourse to these guarantees and support campaigns, traditional sources of concessionary finance are likely to dwindle. The VRA/ECG would not only have to brace themselves to cope with regulatory pressure but additionally face competition from IPPs for funds in international capital markets. These funds are required to finance additional generation, infrastructure improvement and rehabilitation, fuel and other operating costs which are bound to escalate as a result of thermal complementation. IPPs could capture part of the hitherto monopolistic market. Problems also exist with collection of sales made to domestic consumers, particularly in respect of "lifeline" bills.

The government has done well, not to harass the utilities for dividends so that they could deploy their earnings for operations and development activities. Moreover, the government has since 1983 established an Energy Fund to embrace among other special levies on petroleum products and electricity. The National Electrification Fund was also set up in January 1989 to support the local costs of the self help Electrification Project (SHEP) and the District Capitals Electrification Project. Government has also granted appropriate tax exemptions on fuel procured for thermal generation. Government is also collaborating with neighbouring states to implement the West African Gas Pipeline Project linking Nigeria through to La Cote d'Ivoire.

Yet still the challenges faced are enormous. Is the local market sufficiently liquid to meet funding needs? Would market instruments such as index-linked bonds Euro bonds or partnerships with joint venturers for equity investments be the answer? In the face of regulation, would it be possible for tariffs to be revised upwards to reflect higher current costs or long term marginal costs of supply? What about becoming efficient through further cost cutting, fuel strategies? Is the principle of uniform tariffs for electricity supply to the same consumer categories e.g. mines and industries on high voltage supply in jeopardy with the advent of IPPs.

The above issue would have to be addressed and how the utilities can respond to the challenges would determine their resourcefulness. Ghana wishes to learn from India's experience. This Congress is, therefore, timely.

Thanks for your attention and congratulations to the Indian Member committee of the World Energy Council and The Institution of Engineers (India) for organising this Congress. The Chief Executive and Management of VRA extend their felicitations.

ENERGY SCENE IN SOUTH ASIA REGION

G.B.A. Fernando
Secretary, Sri Lanka Member Committee-WEC
Coloumbo (Sri Lanka)

INTRODUCTION

The South Asia region has seven countries viz. India, Pakistan, Nepal, Bangladesh, Bhutan, Sri Lanka and Maldives. In addition Myanmar and Laos Republic have also been considered by WEC under this region. The population structure gives overwhelming majority to India. Bhutan and Maldives are very small countries. The nine countries have wide variety of geographic and topographic locations. Whereas, Maldives and Sri Lanka are small island countries in the Indian ocean, Nepal and Bhutan represented truly cold climate of the eastern Himalayas, where fossil energy resources are not available in a sustained manner. Most of the commercial energy Sources are located in India and Pakistan, and to a small extent in Bangladesh. In the overall situation, the region is a net importer of commercial energy. Six of these nine countries under consideration have coastal belt and are thus exposed to cyclones and other natural calamities.

SOCIAL CHARACTERISTICS

The South Asia Region (except Bhutan) had been under colonial rule for a long time. Independence for the States also posed challenge to individual countries to develop their infrastructure and capabilities. Starting from very low development activities, the countries had to build infrastructure for economic development on one hand and to provide minimum sustenance for its population on the other. The region has over 70% of its population living in the rural areas and substantial numbers facing acute poverty.

All the countries in the region have democratic rule except for Bhutan and Nepal that still have the monarchy.

Demographic Scenario

The total population of the region in 1994 was 1252.3 million (UNDP Human Development Report, 1997) constituting about 25% of global population. This will rise to 1391.5 million in 2000.80% of them reside in the rural areas. The 1960 to 1994, a 34 years growth profile has varying reflections of annual growth rates. The annual population growth rate of Sri Lanka at 1.7% is the lowest and that of Pakistan at 2.9% is the highest. Most of the countries reflect a population growth rate of 2.2 to 2.4%. It is assumed that the efforts made in family planning will considerably bring down the population by the year 2000. Refer Table 4.1.

Table 4.1: Demographic Profile

Sl. No.	*Country*	*Estimated Population (Million)*			*Annual Population growth rate (%)*	
		1960	*1994*	*2000*	*1960-64*	*1994-2000*
1.	Sri Lanka	9.9	17.8	18.8	1.7	1.0
2.	Maldives	0.1	0.2	0.3	2.7	3.5
3.	Myanmar	21.7	44.3	49.3	2.1	1.8
4.	Lao PDR	2.2	4.7	5.7	2.3	3.1
5.	India	442.3	913.5	1,006.8	2.2	1.6
6.	Pakistan	50.0	132.7	156.0	2.9	2.7
7.	Bangladesh	51.4	116.5	128.3	2.4	1.6
8.	Nepal	9.3	20.9	24.3	2.4	2.6
9.	Bhutan	0.9	1.7	2.0	2.1	2.6

India has the highest population of 913.5 million. The second is Pakistan with 132.7 million and the lowest population is of Maldives of 0.20 million.

The change in rural population in these countries over the last 34 years has not shown any major shift. The present rate of urbanisation indicates that in due course of time, say by year 2000, some of these countries will have transferred their significant population to major cities. Urban population as percentage of the total has been reflected between 20 to 27% in 1994. There has been a gradual shift mainly in India where the urban population grew from 18 to 27% and in Pakistan where it grew from 22 to 34% during 1960-94. It is expected that major cities with population over 750,000 will be a serious problem as they will be inhabiting significantly high populations with high population densities. Four of the Mega Cities of the world, namely : Yangoon, Mumbai, Karachi and Dhaka are in this region. Growth rate of these cities has been reflected to be over 4%.

Table 4.2 : Growing Urbanisation

Sl. No.	Country	Urban Population (as % of total)			Urban Population Ann. growth rate (%)	
		1960	1994	2000	1960-64	1994-2000
1.	Sri Lanka	18	22	24	2.4	2.7
2.	Maldives	11	27	28	5.4	4.2
3.	Myanmar	19	26	28	3.1	3.7
4.	Lao PDR	8	21	25	5.3	5.9
5.	India	18	27	29	3.4	3.0
6.	Pakistan	22	34	38	4.3	4.6
7.	Bangladesh	5	18	21	6.3	5.4
8.	Nepal	3	13	17	6.8	6.9
9.	Bhutan	3	6	8	4.7	6.0

It is interesting to note that the population doubling periods for different countries in the South Asian region vary substantially. Whereas Bhutan is expected to double its population by 2020, India is expected to double its population by 2037 and Sri Lanka is expected to double its population in 2065*.

This has also resulted in urban slums created by the migratory population; and semi-urban permanent settlements. Their infiltration into the urban centres also results in a wide variety of fuel mixes appearing in the urban centres. Household use still involves a complex mix of fuelwood, kerosene, electricity and LPG.

The life expectancy is relatively low. The infant mortality rate and children dying before the age of one year is very high. Underweight children under the age of five have been reported to be 50% in South Asia during 1990-96.

Adult illiteracy is very high, specially for females. The female illiteracy in 1995 was reported to be 63% on an average, and that of the male was about 40%.

The human progress in South Asia (only SAARC countries) has been described (August 1996) in the following words by Dr. Mahbub ul-Haq.

"The human progress in South Asia is now beginning to lag behind all other regions, including Sub-Saharan Africa. The over one billion people in the SAARC region are not only amongst the poorest in the world; they lack basic social services of education, health and nutrition. These are 380 million people who are illiterates (of which two-thirds are females); 280 million lack elementary sanitation; and 300 million live in absolute poverty, at the lowest margin of existence. While East Asia has converted their people into a major national asset—through a liberal investment in their education, particularly in technical skills—South Asia has consistently neglected such human Investment. South Korea and Malaysia spend around $130 a year per person on human priority areas of basic education, primary health care, family planning, safe drinking water and nutritional programmes. Contrast this with an expenditure in South Asia of $2 per capita in Bangladesh, $3 in Pakistan and $9 in India. East Asia has also invested a lot in technical and vocational education and in scientific and technological research. That is why South Korea has 46 scientists and technicians per 1000 people, compared with only 3 in India and Pakistan.

At the heart of development strategies of all fast growing economies is massive investment in people and technology. But, this approach is totally missing in South Asia. The adverse effect is reflected in the fact that with the sole exception of Sri Lanka, all other countries fall in the low human development category".

ECONOMIC CONSIDERATIONS

All the countries in the region with the exception of Sri Lanka are classified as low income and low developed countries. In the UNDP 1997 Report, Sri Lanka occupies the highest position amongst the South Asian countries with human development index rank at 94. Most of the other countries occupy position from 131 to 155 out of 175.

The region's economy over the period 1960-94 grew at about 2% in constant 1987 US Dollar terms. As a matter of fact, when converted into real dollar terms, it just matches with the average population growth in the region.

The income disparities have been quoted by UNDP Human Development Report to be alarming, which is also indicative of high rate of poverty. During 1980-94, the poorest 20% in the South Asian Region had an income of US $ 586 in terms of purchasing power parity. Whereas during the same period the richest 20% had an income of US $ 2833 (PPP). If the comparison is to be made between the richest 5% and the poorest 20% when the disparities become still more alarming. These countries also have high level of foreign debt; export raw mineral resources and agricultural produce; and import capital goods in large quantities to build/modernise infrastructure.

Some of the issues that are of major concern in the South Asia Region are :

(i) High rate of poverty and lack of income generating opportunities in the rural areas.

(ii) High incidence of unemployment.

(iii) Dependence on government for economic development and for employment.

(iv) High rate of interest.

(v) High rate of foreign debts and debt service charges.

(vi) High rate of inflation.

Possible Steps to be taken by the Government to reverse trends

Anti-poverty measures supported in the region by different governments are creating an environment wherein people can enjoy a better quality of life. Direct attack on poverty has been the theme of development process and funding by various international agencies in the region after the World Summit on Social Development in June 1995 in Copenhagen. The World Bank, IMF and UNDP have all oriented their programmes so as to offer better opportunities for growth in the rural areas.

Some of the priorities identified for government action in the region are as follows :

(i) Creating enabling environment for development of small scale agriculture, micro enterprises and the services sector will develop centres on which most poor people depend for livelihood.

(ii) Fostering the technological process is the second key priority in the region.

(iii) Reversing environmental decline in marginal regions is the third key priority.

(iv) "Education and Health for All" is the fourth key priority for containing the rural masses in the rural areas.

(v) The last important key priority is the diversification of small cottage industries which can boost rural economy at the resource centres.

ENERGY SECTOR

Overview

The use of primary commercial energy in South Asia increased at an annual rate of 6.8% between 1971 and 1993. Global studies that have been undertaken recently project continued growth during the next several decades. ESCAP (based on IEA study) has given results for the South Asia region for the energy demand using a number of assumptions such as economic growth, energy price, technological progress, consumer behaviour and government policies on energy which need to be treated independently. IEA has given the consumption levels for the year 2000 and also projected two cases of energy demand in the year 2000 and 2010. These studies are, however, focusssing on fossil sources of energy and hydro and geo thermal power. RWEDP has indicated the share of biomass and wood energy in the total consumption of the region will not change substantially. Overall contribution from these sources amount to about 40%. Contribution of non-commercial sources of energy in the South Asia region has been significant and will continue in the similar pattern for a long time.

Very often, energy perspective does not take into consideration the poverty factor. Traditional energy has been central to basic nutrition and health needs of the rural poor. The low income households tend to rely on a significantly different sets of energy carriers than the rich. It is observed that the poor still depend on agro-residues, twigs and branches of the trees which are available free to them. Also, animal dung converted into cakes contribute significantly to the rural energy supply. The better off people then move on to coal, kerosene, electricity or LPG, whichever is available. Such a movement is directly related to the initial cost, cleanliness of the energy source and subsequently the efficiency of use.

Recent studies has established relationship between income and quantity of energy used and income and type of energy source. Two observations are critical :

(i) It has been observed that the total energy consumption per capita in absolute terms increases with the per capita income of the household. This has also been reflected by UNDP. The recent cross country data for 114 countries shows linkage between energy consumption and distribution of income.

(ii) Second, the mix of energy carriers varies with income and its distribution. Whereas in the rural areas in India with larger income transfers due to increasing cereal procurement prices there is a shift from traditional biomass fuels to kerosene and LPG, other countries are still dependent on traditional fuels.

Demand/Markets

The demand projections in South Asia given in Table 4.3.

Table 4.3: Energy Demand—South Asia—Capacity Constraints Case

Levels	*1971*	*1993*	*2000*	*2010*
Primary Energy (MTOE)	*72*	*251*	*355*	*658*
Solids	39	129	153	270
Oil	27	82	128	233
Gas	3	31	57	125
Nuclear	0	2	4	5
Hydro	3	8	12	23
Geo/Others	0	0	1	1
Final Energy	*55*	*162*	*245*	*461*
Solids	25	44	58	93
Oil	23	72	112	207
Gas	2	18	31	74
Electricity	5	28	44	88
Electricity Output (TWh)	*76*	*419*	*650*	*1241*
Solids	32	250	363	710
Oil	6	25	43	65
Gas	4	60	90	178
Nuclear	1	6	14	19
Hydro	33	97	139	268
Geo/Others	0	0	1	1

Source : ESCAP and International Energy Agency (1996).

Resources

(a) Renewable

The renewable sources of energy are mainly in traditional form of biomass which includes fuel wood and agro residues. The other part of the renewable is the new renewable mainly biogas plants, biomass based power systems, solar energy, wind energy, ocean thermal, sea wave power and tidal power. Whereas authentic data on fuelwood is available for most of the countries authentic data on potential of other renewables for all countries is not readily accessible. Therefore, the renewables energy potential, as given out by the government of India, is given in Table 4.4.

Table 4.4 : Renewable Energy Potential in India

Source/Technology	*Potential Availability*	*Potential Exploited*
Biogas Plants	*12Million*	*2 Million*
Biomass-based Power	*17,000 MW*	*Marginal*
Efficient Woodstoves	120 Million	18.5 Million
Solar Energy	5×10 15 Whr/Year	
Small Hydro	10,000 MW	250 MW
Wind Energy	20,000 MW	250 MW
Ocean Thermal	50,000 MW	
Sea Wave Power	20, 000 MW	
Tidal Power	9,000 MW	

Source : National Paper on Energy Scenario in India, 1995, Indian Member Committee, World Energy Council.

(b) Non-Renewable

The energy resource availability, production and consu nption in different countries of basic fuels is as given in Table 4.5, 4.6 and 4.7.

Table 4.5 : Resource Availability (Proved Recoverable Resources) (1993)

Sl. No.	Country	Oil mt	Natural Gas *b cum	Coal mt	Hydro Power MWe	Nuclear Power MWe
1.	Bangladesh	0.5	370	—	N.A.	—
2.	Bhutan	—	—	—	N.A.	—
3.	India	776.0	686	69.947	78.000	1720
4.	Laos	—	—	—	NA	—
5.	Maldives	—	—	—	—	—
6.	Myanmar	6.8	278	2.3	NA	—
7.	Nepal	—	<0.1	<0.1	40.000	—
8.	Pakistan	—	27.0	646	734	725
9.	Sri Lanka	—	—	—	—	—

• Oil and Natural Gas Liquids *mt*-million tonnes, *b cum*-billion cubic metres.

Table 4.6: Energy Production (1993)

Sl. No.	Country	Oil mt.	Natural Gas b cum	Coal mt.	Hydro MWe > 10MWe	Nuclear MWe <10MWe
1.	Bangladesh	0.03	6.16	—	230	—
2.	Bhutan	—	—	—	350	5.66
3.	India	27.03	20.37	263	19.600	243
4.	Laos	—	—	—	230	5
5.	Maldives	—	—	—	—	—
6.	Myanmar	0.72	1.1	0.07	288	—
7.	Nepal	—	—	0.11	234	—
8.	Pakistan	2.94	16.25	3.3	4726	15
9.	Sri Lanka	—	—	—	1137	22.5

Table 4.7: Energy Consumption (1993)

Sl. No.	Country	Oil mt.	Natural Gas b cum	Coal mt.	Hydro Power. MWe	Nuclear Power MWe
1.	Bangladesh	1.1	6.16	—	Same as production	Same as production
2.	Bhutan	—	—	0.02	-do-	-do-
3.	India	54.3	18.6	273.0	-do-	-do-
4.	Laos	—	—	—	-do-	-do-
5.	Maldives	—	—	—	-do-	-do-
6.	Myanmar	0.75	1.1	0.08	-do-	-do-
7.	Nepal	12.0	—	0.11	-do-	-do-
8.	Pakistan	12.0	14.5	4.24	-do-	-do-
9.	Sri Lanka	1.87	—	—	-do-	-do-

Ref : World Energy Resource Handbook

Energy conversion

Energy conversion at the moment is followed in different primary and secondary forms. Coal and biomass are subjected to direct use in the urban as well as in the rural areas. Coal is also being converted into electricity. Most of the power plants on the mainland South Asia still use coal and it is being used in urban and semi-urban townships. Coal in its primary form as well as in its converted form is expected to continue its contribution over the next 25 years as a dominant source of energy in India.

Biomass is traditionally used as fuelwood in cooking and rural energy, and is being converted into gas which is then used for heating or for generation of electricity.

Distribution

Distribution of different forms of energy is mainly still under the control of the government owned energy supply institutions such as State Electricity Boards, and State owned Oil Companies and Coal Companies. Though recent steps have been taken to privatise these sectors, the distribution of electricity is well-organised in different countries. There is also an inter-country link for electricity supply between Bhutan and India through which electricity generated from hydro power is evacuated to India. Under new agreements now signed between India and Nepal and discussed between India and Pakistan the possibility of electricity grids amongst these countries is considered a reality. Distribution of electricity at the user end to households is normally the responsibility of municipalities and rural cooperatives. Privatisation of power sector has been mooted in the region, but the progress is very slow and that is taking place mainly in the electricity supply generation sector. The transmission and distribution sectors have still to be privatised.

Distribution of coal (in India) to thermal power stations and to the industrial and household sectors is well organised. At times, the shortage of coal is temporarily felt because of disturbance in transportation. Transport of LPG and charcoal is also carried out in the same fashion. Whereas large quantities are transported through railways (which is under the government) and smaller quantities are transported through road transport which is in the hands of private companies.

Fuelwood is grown not only in the natural forests but also on private agricultural and non-agricultural barren lands. Some parts of the country have larger production of biomass as compared to others, though the fuelwood consumption is basically uniform with all rural areas on the per capita basis. Also fuelwood supplies to urban areas has to be transported from a long distance. Thimpu (Bhutan) and Kathmandu (Nepal) get fuelwood supplies from a distance as long as 400 kilometres. This does create certain temporary shortages. As different governments have now allowed fuelwood to be grown on private lands it is slowly becoming a traded commodity.

Future Prospects for Energy Supply

(a) Biomass

Energy supply options in South Asia will include the biomass resources for domestic use in a big way whereas all the rural households need biofuels. In Sri Lanka, almost 50% of the households, even in Colombo, use biofuels.

Dependence on fuelwood to the year 2010 is unavoidable. Introduction of high efficiency enduse devices; and higher biomass productivities and larger areas under biomass production reassume that there will be no serious fuelwood shortage till 2010. Localised shortages may be observed, which will be mostly taken care of by fuelwood plantations on government and private lands in the non-forest areas. New initiatives in agro-forestry will promote high productivity and inter-cropping between fuelwood species and agricultural crops. At the same time, increasing urbanisation and inter-fuel substitution will ensure containing fuelwood consumption at present level. A small amount of demand will be met by new and renewable energy sources, such as wind power and solar energy.

(b) Oil and Gas

Oil and Gas will contribute to about 35% and 19% of the total energy with hydro providing about 3.5%. The total requirement for oil is expected to be 233 mtoe and for gas it is 125 mtoe in 2010. The reserve to production ratios of oil in India is about 30 years. The reserve to production ratios in Bangladesh, India and Pakistan are only upto 40 yeaRs. Thus large increases in domestic production cannot be anticipated in South Asia unless major new discoveries are made. The countries have not yet been adequately explored for natural gas and a high priority could encourage increases and exploration both off-shore and on-shore. With limited indigenous capabilities for production, the countries will increase import of oil and gas from outside.

A parallel approach had been suggested for India and Pakistan to import natural gas from the Middle-East and the Central Asia (where several of the countries have very large reserves to production ratios). Suggestions have been made for regional cooperation so as to bring down the cost of transport. India and Pakistan may think in terms of sharing installation of the pipeline from Oman and other Middle-East countries. Such an arrangement was expected to greatly benefit both countries. Natural gas supplied via pipeline was projected to cost about 35% less than LNG in Pakistan and Western or North India. However, the recent shift in approach for gas transport from pipelines to tankers (in liquified form) rules out this possibility. Oman has already announced given up the idea. As the gestation period for the gas pipeline is long and investment high, normally investments are difficult to access.

(c) Coal

Coal and lignite continue to supply in larger share of commercial energy, which will stand at 41% in the year 2010. Heavy dependence of South Asia on Coal is unavoidable as that is a local low cost energy resource suitable for most of the applications. The contribution of coal in south Asia is expected to be 270 mtoe in 2010. Localized shortages of coal and non-availability of high grade coal will call for import of coal in South Asia. Over the past 5 years, India has gradually increased import of low ash content coal for its thermal power plants and this quantity is expected to increase further.

Future Market Changes

With the shift in energy consumption pattern in the rural areas, there will be a substantially higher demand for electricity and gas. How can that be attained? Regional cooperation will have to be promoted for sharing of resources which results in benefit to all stakeholdeRs. An example often quoted is better exploitation of hydro-resources in Nepal and Bhutan and supply of electricity of India, Bangladesh and Myanmar. Such a cooperation can augment electricity supply in the East and North-East India mainly in the rural areas. In this case, the investments can be shared and benefits can be shared on acceptable terms leading to self-sufficiency in the region.

With privatisation of the electricity sector, new power generation capacities are likely to be established, which will, in turn, mean larger energy services. Evaluating in the local context, it will mean more reliable energy supplies at a higher cost as compared to the present erratic and unreliable energy supply at a subsidised low cost. However, the energy market development will be based on true pricing of fuels rather than administered pricing. Development of regional power markets will also ensure higher finance and investment as the financiers will have more than one country to provide guarantee.

INSTITUTIONAL ASPECTS

South Asia region has a strong coordinating group under South Asian Association for Regional coordination, SAARC. It was established in early December 1985, following the first Summit. It is comprised of all the seven South Asian countries (and not Myanmar and Laos). It has been formed to ensure cooperation in economic, social, cultural, educational, and scientific fields and to ensure peace and stability in the region.

It has a programme of coordination in the energy sector focusing especially on renewables in the rural areas. A number of effective steps have been taken in the biomass and other renewable sources of energy. As the fossil fuel energy consumption in rural areas in India alone outstrips all other countries, the responsibility for creating similar energy institutions or facilities also lies with India.

Energy has been a central theme to all the countries in this region in their development planning. It has, therefore, received attention at the central planning level for the whole country. Various international funding agencies like the World Bank, UNDP, FAO, Commonwealth Secretariat and ADB have supported energy infrastructure building. Every programme has given certain important inputs. Besides, the countries in South Asia had the opportunity to discuss their issues under SAARC, G-77 and the DSCAP Energy Programme.

The World Bank has been supporting major energy sector programmes—leading to capacity building. Structural Adjustments has been the major thrust to promote private sector involvement in the region. A point to remember is that World Bank provides only $ 4 b. towards creating appropriate environment for promotion of private sector participation. This fund is global contribution of the World Bank. The total requirement for funding at present is $ 200 b. Thus, World Bank contributes only 2% of global requirement whereas it has the catalytic role it does not in itself fund the project installation in South Asia. GEF Fund (mainly contributed by the World Bank) is to promote environmentally clean technologies. This has really not taken-off so far. However, UNDP has been supporting some other rural energy projects and renewable energy projects. The implementing/executing agencies have been AIT, APDC, ESCAP etc., along with the respective governments.

Most of these programmes were such where capacity building at the national level and the implementation strategy at the local level were promoted. Depending on the priorities of different governments in the regional capacities have been created in terms of energy planning and its linkage with the provincial government for implementation.

ESCAP has been involved in identifying the policy issues that will ensure energy supply in response to the identified needs. At the same time, it is also identifying areas where regional cooperation can lead to energy sufficiency in the region. An example to quote is that ESCAP has suggested regional cooperation between India and Nepal on sharing hydro electric power.

Two major programmes which aided capacity building as well as helped in implementation have been RWEDP and the CHOGM.

RWEDP provides a coordinating mechanism whereby FAO's Regional Wood Energy Development Programme promotes regional cooperation in the countries in Asia and Pacific in the area of wood energy. The RWEDP has helped member countries to learn from each other's best practices in policy making, technology development, manpower planning and programme implementation. A major achievement of RWEDP has been creating consciousness in the member countries to generate authentic data about wood fuels and start planning its use.

CHOGM was a Commonwealth effort launched in 1978 and existed till 1989. This programme dealt directly with the rural energy applications in the region. Under this programme, technology skills and equipment were provided by Indian and Australia in response to the needs identified in the region. In most of the cases, the first time demonstration equipment helped in creating active awareness. Multiplication of the same was then observed wither under respective government's own programmes or under international funding.

RWEDP and ESCAP also work as data bank on energy resources in the region.

India, Pakistan, Bangladesh and Sri Lanka have not only the central government ministries to handle different sectors of energy, but also supporting institutions/

organisations under the government and private sectors to promote large scale support for the energy sector. Nepal focuses on energy through Water and Energy Commissions Secretariat. Bhutan and Maldives have similar focus at the national level.

India has a Energy Framework aimed at meeting energy demand as far as possible from the locally available resources. In the short term, the emphasis is on maximising returns from the assets already created in the energy sector. In the long term, the strategy is to ensure gradual shift from non-renewable to renewable sources of energy with increasing emphasis on demand management, conservation and efficiency in energy production and use. India moved as per the strategy outlined for the Eighth Five Year Plan, i.e. 1992-97. Whereas the private sector and foreign participation in electricity sector moved slowly, various conservation and efficiency improvement steps helped government save 6000 MW equivalent of electricity generation during 1993-1996).

FINANCIAL ISSUES

The South Asia region had serious constraints of financing the energy projects. That is also one of the reasons for slow growth of the energy sector. Persistent fiscal deficit to most of the South Asian countries is indicative of their financial health. Financial constraints on account of balance of payment affected adversely the efforts to supplement additional capacity. Further political uncertainties had also slowed down the development activities.

A key reform considered was opening up of the energy sector to private industry. The private investments provided an opportunity for faster growth of the energy sector, but that also called for financial viability of energy projects. Whereas the responsibility of various governments is to serve the domestic and agriculture sector at a reasonable and affordable energy price, the private investment called for returns that make investments financially attractive. Large number of people in South Asia cannot afford such commercial energy. Therefore, they will not access commercial energy as they cannot pay a price which covers the cost of provision. WEC forum on South Asia held in New Delhi in November 1996 observed : For as long as the electricity tariff in rural India is little more than 3% of total cost of supply, a little more than 25% of fuel cost along, supplies with remain severely inadequate; necessary infrastructural investments will not take place; and the technology and finance desired from other countries and international organisations will not be forthcoming. Innovative financing mechanism were searched in many cases to help bridge the gap between existing situation and the introduction of full market prices. Countries in the region were supported by the World Bank and the International Monetary Fund through long term low interest loan. Due to lack of absorbing capacities or due to administrative bottlenecks such approved and agreed loans were not fully utilised.

Government Funding

Government funding for the energy projects is substantially high in South Asia region. India has been continuously providing about 30% of its budget to the energy sector in the last eight Five Year Plans. Other countries in the region have also been allocating similar amount.

Tariffs

Due to lack of reliable power supply and the social needs of the region, the tariff structure never reflected the actual cost. Power supply different countries. Energy supply cost was administered and fixed by the government and distribution remained a subject of the provincial governments in depended on the application. Whereas the agricultural power supply in India was charged at 0.5 per cent (US) per KWh, the same supply to industry and commercial use was charged at 10 per cent (US) per KWh. However, after the structural reforms in India, there has been a consistent demand to charge the power supply to agriculture at least 1.5 per cent (US) per KWh in the first phase. With the government's permission to generate electricity in the private sector and to transmit the same, it has been observed that power purchase price charged by private producers at about 5 cents (US) per KWh is also acceptable to farmers. The tariff structure does not allow full recovery of the investments in the energy sector and the public utilities are specially in huge losses. This disparity has come under severe criticism from the World Bank and other financing institutions.

RESTRUCTURING OF POWER SECTOR

Three Cent Power in Bangladesh

Bangladesh created history by announcing in January 1998 that it has offered AES Corporation to build, own and operate a 360 ME gas-fired combined cycle plant at Haripur, 12 miles South-East of Dhaka. AES had quoted a tariff of 2.8 cents per unit. That made a big news in Bangladesh, India and Pakistan. The worst effect was obviously in Pakistan as the controversy of their existing IPP at 6.5 cents was already attracting public criticism. However, this news also created a stir in India, giving leverage to those opposing private power production.

Certain facts have now been reflected as to why the kind of figures obtained in Bangladesh may not be possible at other places. The Haripur Project will use natural gas supplied by a Petrobangla subsidiary at almost half the price in India. There were no import duties on equipment as against the 22% customs charged in India. At the same time, Tax Holiday in Bagladesh will be 15 years against 5 in India.

On the financing side, the Haripur Project has support from the World Bank. The tenure is longer and the interest rates lower than those paid by Indian IPPs. The project also has almost absolute guarantee from the government of Bangladesh.

Another important aspect is that a lot of pre-development work was completed by. local authorities, prior to inviting bids. This will not only reduce the development time, but also lower the level of risk perceived by the lenders.

It is difficult to say at this stage, whether the Bangladesh Power Authorities will be able to negotiate a similar PPA with any other IPP.

The controversy of PPAs has cautioned a number of IPPs in India. The demand for reasonable prices has very often been raised. The incident of Bangladesh Power Project make sure demands much more justified.

Some of the examples quoted for the IPPs in this region referred to Indonesia, Thailand and Laos. Most IPPs in Indonesia, signed over the last couple of years have been in the range of 6.5 to 7.5 cents per unit. Even after the currency crisis and the decline of Rupiah, the latest PPA signed in Indonesia for the Tanjung Jai C Power Project quoted a price of 5.73 cents per unit. This despite the fact that the fuel used by the plant would be supplied at a relatively low cost. Similarly recent negotiations between the Thailand buyers and Laotian suppliers have centred on a tariff of 5.7 cents per unit. This source of power is hydro plants. Most of the older contracts in Thailand have been signed at a tariff higher than 6 cents per unit. The IPP Project in Pakistan were signed at a tariff of 6.5 cents per unit. There are also incidents of PPA signed during the last two years in other emerging countries, in the range of 6-8 cents per unit.

The Government of India has announced setting up the Central Electricity Regulatory Commission and corresponding agencies at the State level. The ordinance specifies that the new Commissions should be guided by the following principles when deciding tariffs:

- No consumer to be charged less than fifty per cent of the cost of supply.
- If considered necessary, the State commissions may allow agricultural consumers to pay less than 50 per cent but only for a period of three yeaRs.
- If State Government want to give subsidies to any group of consumers, they must compensate the SEB.
- The tariff should progressively reflect the cost supply.

ISSUES IN REGIONAL COOPERATION

The following six issues have been identified for regional cooperation.

SHARING ENERGY RESOURCES

South Asia region is abundantly endorsed with hydro resources, biomass resources and coal. This can be shared in the region on the market basis. Though most of the organisations and companies in the energy sector are under the administrative control of respective countries. The arrangement for sharing these resources in the region can be established on the basis of market economy.

Sharing Investments in Energy Conversion System

Investments in major projects are sometimes beyond the capacity of the individual countries. For example, hydro resources in Nepal and Bhutan need to attract international financing. They have to provide sufficient guarantees which may not be of total satisfaction to the financing agencies. If a long term agreement is drawn between two or three countries, it will be much more reassuring for the investors to provide finances as the risk for recovery would be substantially low.

Specific projects need to be identified.

Regional Grid

Regional grid in South Asia can be promoted for sharing electricity in the region. Such a regional grid may be shared by India, Nepal Bhutan and Bangladesh in the North-East or by India and Sri Lanka in the South.

Energy Industry Collaboration in the Region

Energy Sector Industry Level collaboration is crucial for rejuvenation of energy supplies (in the final form) to the useRs. Apex Industry Focal Points should take strengthening initiative for regional cooperative programmes by creating infrastructural and manpower skills to promote energy security in the region. This arrangement will support the bilateral and multi-lateral programmes of various governments. This will also provide flexibility and efficiency increase.

Develop a Human Capital Resource

The Institutional capabilities in the region can be shared for building up human skills and promoting Human Resource Development in the energy sector for policy planning, implementation, managements and operation of energy systems.

Regional Financing Mechanism

The financing mechanism to reflect the regional needs and to attract collective financing of the projects can be developed. This may require working at the government level as well as at the industry level. A dedicated Energy Development Fund for South Asia may be promoted by involving international donor agencies.

It may be necessary to have a working group representing interests of the government and of the industry to identify the issues and appropriately address them and facilitate action.

FINANCING OF RENEWABLE ENERGY PROJECTS : HURDLES AND OPPORTUNITIES

L.Y. Bronicki
Chairman, Israel National Committee of the World Energy Council

Renewable Energy Projects are generally small, a few tens of MW at most, and often located in rural areas, where geothermal resources area usually found or where large tracts of land are available for use by solar or wind power generation facilities. These projects are also characterized by high initial capital costs to cover the lifetime supply of "fuel energy". The combination of non-fossil fuel and local rural power generation results in social, economic and technical benefits, which make renewable energy project desirable for emerging markets and developing countries. However the multiple benefits are often ignored as these small projects are subjected to the hurdles of long term financing.

HURDLES TO IMPLEMENTATION OF RENEWABLE PROJECTS

Financing Barriers

Commercial Financing barriers results from the higher up-front costs of renewable energy projects. The cost of fossil fuel power includes the power plant capital cost and a per unit fuel cost, while in renewable projects the capital cost includes the equipment for covering natural energy. As a result the capital cost for a renewable project is much greater than the comparable cost for a fossil fuel project of the same size. Commercial institutions view this as increasing the project's financial risk profile.

Credit Risk barriers, are particularly prevalent in renewable energy projects since these projects are most desirable in developing countries. As a result countries which can benefit the most from renewable projects are least attractive to the financing institutions.

Institutional barriers may exist in developing countries due to national policies under which fossil fuel power is sold at subsidized prices. As a result the actual cost of renewable energy is often compared with the low price of subsidized electricity.

Multilateral Agencies Financing for Emerging Markets

Multilateral agencies are often the only source available for financing renewable projects in emerging markets.

Renewable Energy projects are small compared to fossil fuel plants and cannot receive the allocation of manpower from the agencies, which may be focused on large power projects.

Financing Cost procedures and time tables are geared for large national projects. This imposes costly time consuming hurdles on the smaller renewable energy projects.

Statutory Limitations on the share of the financing that single agencies can subscribe to, usually 30 to 35%, necessitates that two or more agencies must work together, each with its own specific procedures and with associated complications and delays.

Standards and Costs of Review

Review Standards for small renewable projects, utilizing innovative technologies, are often more stringent and time consuming than for well defined large fossil fuel plants.

Financing Expenses, especially if more than one multilateral agency is involved, are often the same, in total amount, as for the large fossil fuel projects, and are thus grossly disproportionate in size for small renewable projects.

Time Delays in the extensive financing process may affect national goals and in fact may weaken the project's feasibility.

ACCELERATING RENEWABLE PROJECT IMPLEMENTATION

The implementation of small renewable energy projects can be accelerated by reducing the expenses and time delays required to arrange and close the financing for the projects.

Dedicated Fast Track Process for "One Stop" Financing

Financing of renewable projects should be processed on a FAST TRACK, utilizing experience with similar projects, standardized procedures and increased reliance on the history of the project participants.

Agency financing Requirements should be identified at the beginning of the review period to keep the financing process on a smooth fast track.

The Review Period for the project through financing approval, should be targeted at six months. This period is sufficient for a diligent and thorough review.

The Financing Participants, so long as the 30 to 35% single agency statutory limitations are in force, should choose as TEAM LEADER, who will represent the needs of the participants without having to coordinate the time schedules. This will keep the financing process on schedule and control its costs.

Agency Teams should be assigned to the projects from inception through project review and financial closing. The shorter financing period will avoid the need to change teams during the process because of the needs of other larger projects. This makes for a smoother financing process.

Standard Agreements, such as were used in the USA for small power projects, should be developed and specifically tailored for financing of renewable energy projects.

The Closing Schedule, including costs and timetable should be agreed upon during the review period.

Accelerated Implementation Benefits

The reduction of the financing hurdles which small renewable energy projects must endure will accelerate project implementation and produce the following benefits :

- *Project Costs* will be decreased, thus decreasing the cost of power production. This will make renewable energy more cost effective and encourage more such projects.
- *Electrical Power* will be produced earlier and at lower cost to accelerate local and regional economic development.
- *Developer Resources* may be focused on technology improvements and project enhancements. This will provide for more efficient and cost effective future projects.
- *More Renewable Projects* will be developed, when and where needed.

ENERGY SCENARIO IN INDIA

K. N. Sinha,
Member (Planning), Central Electricity Authority
New Delhi

ECONOMY AND ENERGY

Indian Economy

India, covers an area of 328 million hectares and had an estimated population of over 965 million in 1997. Over 25% of the population lives in urban areas. The population of four metropolitan cities of Delhi, Bombay, Chennai and Calcutta is about 5% of the total population.

India's economic growth over the past 50 years since independence has been commendable in many ways. Starting with a narrow infrastructure base and financial constraints, India has developed into a leading technological country. Its scientific manpower reckoned to be the third largest in the world, has built India's capabilities in the frontier areas of science and technology. Moving from an agricultural economy, India is now well on her way to become a major industrial power. The initial stages of development focussed attention on building core sectors and education and training facilities in science and technology.

India has developed considerable capability and technological skills as a result of the early policy of building a strong base of heavy industry within the public sector. Liberalising its economy in 1991, India created an opportunity for public sector organisations to improve management skills, technological competence and sharpen its performing cutting edge. Privatisation and dereservation are beginning to show their impacts. Adoption of the advanced technologies, service facilities and management practices coming in with foreign investment is beginning to transform the scene. Simultaneously, the private and foreign investors have been invited to be partners in sharing responsibilities and profits and benefits by fully utilizing the investment opportunities in the country. Public and Private sector industries now compete on an equal footing.

India has moved carefully in introducing financial and labour reforms that allows maximum exploitation of market economy as some of them do not respond to the societal demands of the poor and agricultural population.

The changing economic environment is opening up opportunities for skill and initiative and promoting a number of non-resident Indians to return to India and reverse the trend of brain drain.

India is now looking outwards and thinking globally. The old notion of self-sufficiency and import substitution is becoming broad-based to promote more outward looking impulses that could impart new dynamism to the economy. Whereas the country is exploring *"financial investments"* and *"capital"* from other countries, it is providing *"intellectual capital"* to the global markets in the form of trained IT professionals and software experts as well as in other high tech areas of engineering, medicine and basic sciences.

The developments in the last few years have sent a strong signal that the economic reforms have come to stay. The *'pace'* and *'pattern'* may vary with any political changes, but the broad thrust of liberalisation will not be reversed. The unleashing of the economy is generating its own momentum. Economics cannot be isolated from political decision making. The 1998 elections have shown that India has made a smooth transition from single party dominance to coalition governance with a tilt towards greater cooperative federalism and this could serve as a valuable corrective to earlier more centralising trends and could be a stabilising factor for the market economy in India.

The industrial growth has accelerated, service sector is picking up and exports are responding to the international markets. High rate of growth during 1992-97 exposed the infrastructural bottlenecks, especially in the power sector which is the prime mover for all growth and development.

The above mentioned elements have resulted in accelerating the growth of the economy. The average rate of growth of the economy has been 6.8 per cent during the Eighth Five Year Plan Period (1st April, 1992 to 31st March, 1997). During the terminal year (1996-97) of the Plan, the GDP growth has been 7.5%. The agriculture and allied sector growth was at 7.9% (with the annual foodgrain production reaching 199 million tonnes); the industrial sector growth was at 7.1%; and the service sector (comprising trade, hotels, transport and communications, financial services, business and public administration) growth was at 8.1%.

Also, during the financial year 1996-97, the rate of gross domestic savings was 26.1% of the GDP, which is considered quite high.

Energy Sector Growth

The rate of growth of consumption of commercial energy has been about 5.5% per annum during the last five decades.

The elasticity of total primary energy consumption for the period 1953-96 with respect to Gross Domestic Product (GDP), at constant price, works out to 0.78 per cent; as compared to 1.25 per cent for commercial energy. The elasticity of total final energy consumption works out to 0.67 as compared to 1.14 for final commercial energy consumption for the same period.

COMMERCIAL ENERGY RESOURCES

Primary Energy Resources and Regional Distribution

Coal and hydroelectric resources are the mainstay for India's primary resource support for secondary energy purposes. Oil and natural gas resources of the country are limited. Nuclear resources, on the other hand, are modest. The coal resources are of poor quality having high ash content. Besides the above, the distribution of the basic resources i.e. coal and hydroelectric resources in the various regions are rather skewed. While the eastern region accounts for 70% of the total coal resources, the western region accounts for over 70% of the hydrocarbon reserves and over 70% of the total hydro potential of the country is concentrated in the northern and north-eastern region. The southern region which has only 6% of the coal reserves and 10% of the hydroelectric potential, possesses most of the lignite deposits of the country.

The regional distribution of the primary commercial energy resources are given in Table 6.1.

Table : 6.1 : Regional Distribution of Primary Commercial Energy Resources

Regional	*Coal*	*Lignite*	*Crude Oil*	*Natural Gas*	*Hydro Power*	
					Potential at 60%	*Energy*
	(BMT)	*(BMT)*	*(MMT)*	*(BCM)*	*LF (MW)*	*(TWH)*
Northern	1.1	1.1	—	4	30135	225.00
Western	48.2	0.5	584	497	5679	31.40
Southern	13.1	25.9	—	—	10763	61.80
Eastern	141.4	—	—	—	5590	42.50
North-Eastern	0.9	—	148	159	31857	239.30
ALL INDIA	204.7	27.5	732	660	84044	600.00

BMT=Billion Metric Tonnes. MMT = Million Metric Tonnes. BCM = Billion Cubic Metre.

Non-conventional Energy Sources

The country is endowed with a large potential of non-conventional energy sources. The potential availability and extent of exploitation as on March 31, 1998 are given in Table 6.2 below :

Table 6.2 :Renewable Energy Potential In India

Source/Technology	*Potential Availability*		*Potential Exploited*	
Biogas Plants	12	Million	2.713	Million
Biomass-based Power	17,000	MW	114	MW
Efficient Woodstoves	120	Million	28.5	Million
Solar Energy	5 × 1015	Whr/Year	32	MW SPV
			4,20,000	Sq.Mtrs. SWH
			4,57,000	SC
Small Hydro	10,000	MW	155	MW
Wind Energy	20,000	MW	970	MW
Energy Recovery from				
Urban and Industrial Wastes	—		4.75	MW
Ocean Thermal	50,000	MW	—	
Sea Wave Power	20,000	MW	—	
Tidal Power	9,000	MW	—	

SPV = Solar Photovoltaic. SWH = Solar Water Heater. SC = Solar Cookers.

Traditional Non-commercial Primary Energy Sources

Fuelwood is an important source of energy in the rural and, to some extent, in the urban households. The total forest area in the country has been assessed at nearly 75 million hectares. The growth in the demand for wood for industry, construction and for domestic use for cooking, etc. has exceeded the annual incremental availability of wood from local sources and social forestry schemes. This has resulted in the gradual denudation of tree cover and consequent adverse impact on the environment.

Fuelwood is supplemented by dung and crop residues in meeting domestic energy needs in the rural areas. The annual availability of wet dung is estimated to be about 960 mt. Greater use of dung cake as a source of energy reduces the availability of dung as a valuable organic manure. Total availability of crop residues is estimated to be 450-500 mt. The draught animal population in the country has been estimated at 70 million and animal energy continues to be used in agricultural operations and for rural transportation.

TRENDS IN ENERGY PRODUCTION, SUPPLY AND CONSUMPTION

Indigenous Primary Energy Production

India's energy requirements are met from traditional as well as modern forms of energy. The distinctive characteristic of the Indian energy consumption pattern is the continued co-existence of non-commercial fuels like fuel wood, agricultural waste and cow-dung cakes and animate form of energy, in conjunction with the commercial energy forms. The share of traditional forms of fuels is diminishing and still contribute to over 34% of the annual energy consumption. The usage is largely restricted to domestic fuels in semi-urban and rural areas at very low efficiencies.

Table: 6.3 :Trends in Indigenous Production of Commercial Primary Energy

Year	*Coal mt*	*Lignite mt.*	*Crude Oil mt*	*Natural Gas bcum*	*Hydro TWh*	*Nuclear TWh*	*Wind Power TWh*
1953-54	33.0	—	0.26	—	2.52	—	—
1960-61	55.67	0.05	0.45	—	7.84	—	—
1970-71	72.95	3.39	6.82	1.4	25.25	2.42	—
1980-81	114.1	4.80	10.51	2.4	46.54	3.00	—
1990-91	211.73	14.07	33.02	18.0	71.66	6.14	0.03
1996-97	288.65	22.54	33.87	22.9	68.63	9.01	0.85

Note : TWh = 10^{12} Watt hours.

Energy Imports

Energy imports are largely in the form of crude oil and oil products. The import/export of other commercial energy forms, viz, coal and secondary electricity, has been rather limited and restricted. While the import/export of electricity is limited to neighbouring countries, import of coal has been limited to superior quality for use in selected sectors and selected places in the steel and power industry.

The import of crude oil and oil products in the year 1996-97 has been 33.90 mt and 20.52 mt (crude equivalent) respectively.

The import of coal in the year 1996-97 has been 10 mt.

Primary Energy Supply

Overview

The total supply of energy (both of commercial and non-commercial forms excluding draught animal power) increased from 89.6 mtoe in 1953-54 to about 368 mtoe in 1996-97. The share of non-commercial fuel supply declined from 74% in 1950-51 to about 34% in 1996-97. Fuelwood continues to maintain its primacy in the non-commercial energy supply (largely in the household sector). Changes in the pattern of primary energy supply over the past four and half decades can be seen in Table 6.4.

Table : 6.4 :Changes in the Pattern of Primary Energy Supply (mtoe)

	Production Years					
	1953-54	*1960-61*	*1970-71*	*1980-81*	*1990-91*	*1996-97*
Commercial Primary Energy						
Coal	23.62	35.64	36.48	56.96	94.68	122.79
Lignite	—	0.01	0.81	1.23	3.34	5.04
Crude Oil	0.19	0.46	7.01	10.79	33.92	33.87
Natural Gas	—	—	0.60	1.41	11.73	20.05
Hydro Power	0.24	0.67	2.17	4.00	6.16	5.90
Nuclear Power	—	—	0.63	0.78	1.60	2.35
Wind Power	—	—	—	—	—	0.07
TOTAL	**24.05**	**36.78**	**47.67**	**75.19**	**151.43**	**190.30**
Net Imports	2.20	6.04	12.66	24.63	31.69	61.63
St. Changes (-)	0.24	2.87	0.69	3.80	5.37	2.38
Intl. Bunkers (-)	0.53	2.50	0.24	0.21	0.14	0.17
Total Commercial Energy Supply	**25.48**	**39.45**	**59.40**	**95.81**	**177.61**	**249.38**
Non-Commercial Primary Energy Supply	**64.13**	**74.38**	**86.72**	**108.48**	**122.07**	**127.51**
Total Primary Energy Supply	**89.61**	**103.83**	**146.12**	**204.29**	**299.68**	**373.94**

The consumption of non-commercial fuels in the year 1996-97 is estimated at 173 mt of fuelwood, 95 mt of dung cakes and 53 mt of agro waste.

The salient features of the energy sub-sectors are described below.

Coal

Coal and lignite (also called brown coal) are the major constituents of primary energy resource.

The demand for coal has been met mostly from indigenous sources. In the year 1996-97, the import of coal accounted for about 10 mt (about 3% of the annual coal production).

Coal is transported from pit head to consumption points, through road and rail. Coal movements by railways presently (1996-97) constitute about 60% of the annual coal production. Road transport is used extensively for coal transport over short distances.

Indian coals are generally characterised by high ash content. Power grade coals are generally despatched and used as Run of Mine (ROM) coal. Coal beneficiation for power grade is gradually picking up, to meet environmental standards. The present annual capacity of coal beneficiation, including coking and non-coking coal, is 10% of the annual coal production.

The coal production in the year 1996-97 has been 289 mt; out of this about 70% has been supplied to the power sector. The next major coal consuming sector is the steel sector needing about 35 mt. Cement needs 10 mt. The remaining is supplied to small industries and domestic sector.

The lignite production in the year 1996-97 was 22.5 million tonnes. Lignite production in the country is limited and restricted largely to Southern India in Tamil Nadu, with some small capacity in Gujarat in Western India, and in Rajasthan in Northern India. Some lignite is also used for production of fertilizers and preparation of lignite briquettes for use in the domestic and industrial sectors.

Coal production was largely in the private sector till its nationalisation in 1973 after which major coal producing companies (except coal for captive purposes) have been in the public sector. The sector was opened to private participation only recently.

Oil and Gas

Oil demand in India was being met mostly through imports till mid-eighties when indigenous production picked up as a result of large discoveries of crude oil in the offshore regions. The annual production of oil from indigenous sources in 1996-97 has been 35.5 mt out of which 62% was contributed by off-shore production. Natural gas production has been 20.0 mcum in the same year.

Refinery capacity for treating crude oil expanded to match the demand for petroleum products. Presently, there are 13 refineries with a throughput capacity of 62 mt. The largest refinery in the country is located at Koyali along the West Coast in Gujarat with a throughput capacity of 10 mt.

India's dependence on import of crude oil and products is on the rise. The present self-sufficiency in meeting crude oil demand is about 45% to 50%.

Electricity

Electricity has been the preferred form of energy with its comsumption registering an annual compound growth of 9-10% over the last five decades, compared to the growth of overall energy consumption at the rate of 5-6%. Whereas the annual consumption of electricity in the industry has declined from 69% in 1960-61 to about 43% in 1990-91 (the annual reduction is also due to power shortages and their impact, mainly on the industrial sector), the share of consumption in the agricultural sector has increased from 8% in 1960-61 to over 25% in the year 1996-97. This was mainly due to large emphasis on lift irrigation and also substitution of oil by electricity in lift irrigation. The household sector also registered increase in the annual consumption levels with its share increasing from about 11% in 1960-61 to about 19% in 1996-97.

The Electricity Supply Act of 1948, sought to reorient and rationalise the electricity supply industry through structural and functional means. This resulted in systematic growth of electricity generation.

The installed generating capacity increased from 1722 MW in 1951 to over 86,000 MW by March, 1997. The annual electricity generation rose to 395 TWh in 1996-97in the public utilities. The generation in the non-utilities is estimated at 42 TWh during the year.

Of the commercial energy forms, coal has been the basic primary commercial energy source. Through the process of planned development over the last five decades, a significant growth has been achieved in the indigenous production of primary energy.Trends in indigenous production of commercial primary energy are given in Table 6.4. (It is assumed that the energy produced is consumed and therefore the figures all refer to consumption).

Energy Consumption

The share of commercial energy in the final energy consumption increased from 28.4 to 65.9% during this period, while the same for non-commercial energy sources has declined from 71.6% to 34.1%. (see Table 6.5).

ENERGY CONSERVATION PROGRAMME

Overview

Energy management and conservation, as a concept, had its origin in the post-oil shock period. Till then, attention was paid only on augmenting the commercial energy supply. After the two oil shocks there has been growing realisation about the scope for energy conservation and its need in long term sustainability of meeting the energy demands. Initially, attention was paid to energy conservation in the production and management of commercial forms, viz, coal, oil and electricity. Even among commercial fuels, oil conservation had the initial focus and gained momentum with the setting up of Petroleum Conservation Action Group in 1976 which was later converted into Petroleum Conservation Research Association.

Table : 6.5:Changes in Pattern of Energy Consumption (in %)

Year	*Main Energy Form*		*Break-up of Commercial Energy Consumption*				
	Non-Comml.	*Comml.*	*Coal*	*Oil*	*Natural Gas*	*Electricity*	*Total*
1953-54	71.6	28.4	80.1	16.7	0.0	3.2	100.0
1960-61	65.3	34.7	75.3	19.9	0.0	4.8	100.0
1970-71	59.4	40.6	56.1	34.1	0.6	9.2	100.0
1980-81	53.1	46.9	47.9	40.3	1.1	10.7	100.0
1990-91	40.7	59.3	35.9	43.6	5.5	15.0	100.0
1996-97	34.1	65.9	29.3	46.8	6.9	17.0	100.0

The shift in the relative share of the final commercial energy consumption in different sectors is given in Table 6.6.

The emphasis on oil conservation had two basic objectives :

(i) Substitution of oil by all other forms in all the sectors of consumption; and

(ii) Conservation of oil in areas where substitution by other forms of energy is economically not viable.

The programmes for oil conservation are supported by institutional back-up for improvement of efficiencies of fuel use which included alternative fuel use etc. Over a period of time, conservation expanded and encompassed other commercial energy forms such as electricity and coal sectors. By mid-eighties, action plans and programmes were

Table 6.6 : Percentage Share in Final Commercial Energy Consumption by Different Sectors (in %).

	Percentage Share in Consumption					
Sector	*1953-54*	*1960-61*	*1970-71*	*1980-81*	*1990-91*	*1994-95*
Industry	39.9	41.5	49.6	49.6	44.7	42.1
Transport	43.7	40.2	28.2	23.5	19.8	20.8
Household	9.6	9.7	12.2	9.7	11.8	12.3
Commercial	0.5	0.3	1.2	0.9	1.1	1.2
Agriculture	1.4	1.6	2.7	6.3	8.2	9.3
Feedstocks	0.5	0.3.	4.2	5.6	10.4	10.3
Non-energy use	2.9	4.7	1.5	4.0	3.1	3.0
Others	1.5	1.7	0.4	0.4	0.9	1.0
TOTAL	100	100	100	100	100	100

being made for each fuel and for each sector of consumption. In the oil sector, attention was focussed on the conservation of energy in transport, agriculture and industry; while in the electricity sector it was agriculture and industry which initially attracted the attention.

By early nineties, conservation programme attained heightened dimension and a National Action Plan for Conservation of Energy was developed.

Action Programmes — Targets and Achievements

Oil Sector

The oil sector planned for conserving 5 mt of oil during 1992-97 through efficiency improvement and conservation measures. The programme in the oil sector for conservation includes :

(i) Awareness Programmes for oil conservation leading to conservation of products and protection for environment.

(ii) Programmes for substitution of petroleum products by alternative sources of energy including new and renewable sources of energy.

(iii) R&D and Pilot Programmes for inception and promotion of fuel-efficient green technologies, equipment and devices.

(iv) Training and Technical Support for promoting efficiency in the oil use.

Industry, transport, agriculture and domestic sectors were the target groups with linkages established for each sector with organisations sharing similar objectives. Financial incentives are provided for arranging mandatory energy audits. These and other programmes created an impact and resulted in an annual saving of 1.36 mt of oil by 1997-98. In the next five years, it has been targeted to achieve conservation of 2.7 mt through a variety of programmes.

Electricity Sector

A National Energy Efficiency Programme (NEEP) was envisaged for the electricity sector. It aimed at saving of 5,000 MW of additional installed generation capacity during the five-year period of 1992-97. The supply side measures were to result in avoiding 2,250 MW, and the demand side measures were to result in avoiding 2,750 MW.

The following have been the targets and achievements on the supply side :

	Target	Achievement
(a) Through efficiency improvement of generating stations with capacity utilisation and reduction in auxilliary consumption.	1350	2135
(b) Through reduction in Transmission and Distribution losses.	900	1150
TOTAL	**2250**	**3285**

On the supply side, renovation and modernisation programme was taken up as a planned activity with a targeted number of 44 thermal power stations with a total capacity of over 20,000 MW. The programme resulted in an additional annual generation of 7864 m KWh per year, which represents increase of 100 MW peaking capacity besides the extension of the life of the units.

In the electricity sector, a target of conservation in the consumption of oil was set at 0.3 mt for the five year period as against which there has been an achievement of 1.4 mt by the fourth year.

On the demand side, a saving of 2,750 MW was targeted. There has been substantial reduction in the specific energy consumption in :

(i) some of the industrial sub-sectors like cement, paper, steel;

(ii) agriculture sector through pump set rectification programme; and

(iii) domestic sector through deployment of energy efficient lighting systems.

For the next five years, a similar Action Programme is proposed. In addition to planning for better capacity utilisation, other management measures include compulsory audit for power stations and industries, with improvement of standards for the specific consumption levels in the end use sectors are proposed.

Coal Sector

Conservation in coal sector has been in the areas of (i) conserving electrical power through a combination of management and other techniques and (ii) through technical support, in coal mining as well as coal washing. The management measures include :

— Maintenance of the plant and machinery through adherence to accepted O&M techniques.

— Incentives for energy conservation.

The technical measures include maintenance of the desired levels of power factor by providing suitable capacitor banks in the electrical system.

ENERGY DEVELOPMENT PROGRAMME FOR THE SHORT TERM

Coal

The next Five Year Plan is under finalisation. However, indications are that the programme for coal sector includes mine development to increase the annual production capacity of coal from the existing level of about 289 million tonnes to reach 371 mt in the year 2001-02.

Besides the increase in physical output the programme for the next five years includes intensified detailed exploration efforts, proving additional coal resources, improving the coal quality and consistency through coal washeries for coking coal and beneficiation of non-coking coal being supplied to power plants situated at distances beyond 1000 kilometers. Other measures proposed include environmental measures, improved coordination for coal transportation and enhanced R&D thrust in the areas of productivity, mine safety and coal beneficiation.

Oil and Natural Gas

The programme for the oil sector includes augmenting the domestic crude production to reach 37 mt and that for gas production to reach 31 bcum in the year 2002. Oil demand is expected to reach 113 mt in the year 2001-02. The annual import of crude oil and oil products is expected to reach 76 mtoe in the year 2001-02. The import dependence in respect of crude oil to meet the demand of petroleum products for the year 2001-02 is likely to go up to 70%. In that context, the following areas are being given attention :

(i) Improving the reservoir management and enhancing oil recovery;

(ii) Rationalisation of pricing structure for POL products in the domestic market in order to reduce demand;

(iii) Creation of additional refining capacities and marketing and distribution facilities;

(iv) Exploring the possibility of substitution of POL products by LNG.

Electric Power

The gross energy generation from utilities in the year 1996-97 has been 395 TWh, indicating an average annual growth of 6.6% over the past five years. The programme for the next five years involves addition of over 40,000 MW of capacity with expansion of associated transmission systems and distribution networks.

The power programme includes participation in the capacity addition programme through private entrepreneurial support, emphasis on renovation and modernisation of existing facilities, rationalisation of mismatch in the transmission and distribution support, reduction in the regional disparities in the electricity availability, reduction in the average gestation time per megawatt of addition to installed generating capacity, better fuel management with targeted reduction in the specific fuel consumption levels,and encouragement to captive power generation including cogeneration.

Nuclear Power

The nuclear power programme for the next five years includes modest addition to the physical capabilities in the nuclear power generation; thrust on performance of nuclear power stations; and upgradation of capabilities for higher capacity addition programme based on indigenous capability.

Non-conventional Energy Sources

The importance of renewable energy has significantly increased in recent years because of growing environmental concerns with fossil fuel based energy and the urgency of promoting sustainable energy development. Renewable energy also has a particularly critical role to play in meeting the growing energy needs in the rural areas. Furthermore, the large scale development and utilisation of renewable energy sources would also lead to the strengthening of energy security by reducing the import bill on fossil fuels.

ENERGY POLICY

Strategic Elements of Energy Policy

In evolving strategies and programmes for the development of the energy sector, the basic objective has been acceleration of energy production and supplies in the most convenient and useable forms and also with emphasis on self-reliance both in terms of primary resource support and technology development. Certain basic policy prescriptions were conceived from the inception of the planned development of the energy economy. These include :

(i) Conservation of limited coking coal resources and confining its use exclusively for metallurgical purposes in the steel industry;

(ii) Reliance on indigenous coal for thermal power generation in steam power plants and near virtual restriction on the use of oil products for power generation;

(iii) Emphasis on development of hydro-electric power to optimise the power development.

Energy strategies adopted from time to time are preceded by detailed study/studies by expert groups. The first such study was carried out in 1961, under aegis of the *"Energy Survey of India Committee"*, with a comprehensive survey of energy resources of the country which enabled formulation of a pragmatic energy strategy during the sixties. During the period between the two oil shocks (1973-79), the energy situation was reviewed twice — once under the aegis of the Fuel Policy Committee in 1974, and again under the aegis of the Energy Policy Committee in 1979. In 1976, a comprehensive energy policy was formulated which, underlined *inter-alia* :

— The importance of coal as the principal source of energy;

— The need for reducing reliance on imported oil;

— The need for substitution of oil by any and all other forms of energy, wherever technically and economically feasible;

— The need for developing new and renewable energy sources with appropriate levels of efficiency in use together with progressive improvements in the efficiency of their use.

Energy strategies are outlined at regular intervals and at the time of formation of the Five Year Plan programmes for the economy, taking into account the national and international developments in the energy economy and their impact on the economic end uses. The elements of energy strategy considered in the context of formulating the draft for the next Five Year Plan, include the following major elements :

— Improving the efficiency of use in the existing installations through technology and management back-up and renovation programmes, efficiency improvements in the production systems and reduction in transformation losses, including those in traditional forms of energy resources;

— Adopting energy efficient technologies in the supply and end use systems;

— In-building and internalising standards for environmental efficiency in the energy production and supply processes and systems;

— Rationalising energy pricing in order that real costs are reflected in the revenue making processes;

— Strengthening and reforming the institutional structures in the energy sector in order to promote the economic policies in the new and adapted atmosphere;

— Promoting R&D in the energy sector to integrate economically the elements of environmental benignness, cost sustainability in the commercialisation of the energy technologies, including those meant for substitution of depletable energy sources by renewable resources and improved utility and convenience in the secondary energy use.

ENERGY SECTOR INVESTMENT

Investments in Different Five Year Plans

Energy sector development since independence has been based on government support and public fund allocation. Electricity sector was taken over by the government in view of its direct impact on industrial and agricultural growth. Subsequently, the coal and oil sectors were also nationalised to meet the social and economic objectives of the country.

There has been priority channelisation of public investments to the energy sector programmes. Every 5 Year Plan allocated upto 30% of its funds for the energy sector; electricity sector receiving allocation support to the extent of 20% of the outlay. Average annual investment in the energy sector was Rs. 772 million per year during the five year period 1951-56 which went up to Rs. 230 billion per year during the period 1992-97. The investment trends/plan expenditure during the five year periods from 1951-56 to 1992-97, in absolute figures and the share of energy sector investments in the overall annual public sector investments/outlays is given in the Table 6.7.

Table 6.7 : Plan Expenditure and the Share of Energy Sector

Plan	*Period*	*Plan Expend. (Rs.billion)*	*Share of Energy Sector (%)*				
			Power	*Oil/Gas*	*Coal*	*NRSE*	*Total*
First	1951-56	19.6	—	—	—	—	19.7
Second	1956-61	46.7	9.7	0.8	1.9	—	12.4
Third	1961-66	85.8	14.6	2.6	1.3	—	18.5
Fourth	1969-74	157.8	18.6	1.9	0.7	—	21.2
Fifth	1974-79	394.3	18.7	3.6	2.9	—	25.2
Sixth	1980-85	1092.9	16.7	7.8	3.5	0.1	28.1
Seventh	1985-90	2202.2	17.4	7.3	3.2	0.3	28.2
Eighth*	1992-97	4341.0	18.4	5.5	2.4	0.2	26.5

* *Outlay*

Energy Sector Investment in Post 1992 Fra

After the liberalisation (in 1991) and with the opening up of economy to private sector, there has been reduction in the public investment to energy sub-sectors in the hope that the private sector will provide the necessary support. Private sector investment was however enough to meet financial needs of the power and coal sub-sectors. As a consequence, during the five year period covering 1992-97, the investment in the power sector has been 2% of GDP as against 2.4% earlier during the Sixth Plan (1980-85). Similarly, investments in coal declined from 0.5% of GDP in the Sixth Plan (1980-85) to 0.3% of GDP in post 1992. On the other hand, investments in oil and gas exploration, oil refining and gas supply increased from 0.7% of GDP in the Sixth Plan (1980-85) to 1.2% of GDP in post 1992.

More than 90% of the investment in the energy sector, till the liberalisation period, has been under public sector ownership. After liberalisation, public expenditure has been relocated to focus attention on social sectors. This together with the lack of the required levels of private sector support, during the initial period of liberalisation led to energy shortages.

The government will continue to play an important role in the energy sector projects. In that context, following issues deserve attention :

(i) Financial guarantee to domestic Developmental Financial Institutions (DFI) through the provision of sovereign guarantees, equity capital and preferred tax requirement.

(ii) Promotion of growth of Domestic Institutional Investors by the government by encouraging higher growth rate of institutional assets.

(iii) Promotion of various other options such as : shift of energy investments from capital to current account by importing energy goods and services; shifting investments to non-energy sectors and privatisation.

(iv) Development assistance and soft term loans from international funding agencies.

(v) Involvement of multinational companies in joint ventures.

(vi) Participation by public and private companies and employees.

(vii) Leasing of energy equipment.

(viii) Institutional development.

ENERGY SECTOR REFORMS

Need for Reforms

There has been substantial rise in the specific capital cost of energy installations. In addition to the above, the energy sector has to provide infrastructure support for the socio-economic programmes.

The policy to promote private participation in the energy sector programmes is guided by the need for additional investment. Environmental management, protection and conservation of the natural resources, has also emerged as a key priority in the reform process in the energy sector in functional terms. Initially started with the power sector the programme of private sector participation has been extended to other energy sub-sectors. Liberalisation in the energy sector was initiated at a time when global change for deregulation and liberalisation was active. India joined the global market in actively soliciting private proposals for development of energy sector programmes.

Different categories of activities for private participation include the following:

(i) setting up of the electricity generation stations;

(ii) coal mine development; and

(iii) exploration, proving and development of oil and natural gas wells etc.

Functional and structural reforms are set in motion in all the energy sub-sectors and the strategic elements of the reforms include the following :

Power Sector Reforms

Functional Reforms (Investment Promotion)

Reforms in the power sector were initiated in 1991. In order to make the climate for investment congenial to the entrepreneurial measures for deregulation and devolution of power to utilities, direct negotiation with the entrepreneurs has been encouraged. Regulations relating to fuel supply have been liberalised so as to attract a variety of fuels to become supportive to power generation based on economy of supply. Legislative measures were effected to support congeniality in the commercial climate in order to make tariffs acceptable to the consumer and encouraging the entrepreneurs. The clearance procedures have been rationalised. Time for techno-economic clearance by CEA has been reduced to 5-6 months. Project clearance based on competitive bidding is made mandatory and the methods for clearance are being rationalised.

Structural Reforms

In structural terms, power sector utilities are endeavouring to unbundle the existing vertically integrated monopoly and separate generation,transmission and distribution activities for rationalising their management. Whereas efforts are made to maintain conceptual uniformity in the unbundling, the multi-ownership for generation and distribution and continuation of government monopoly in the transmission would be the near term approach in the initial stage of reforms. Utilities are currently debating the possible structures. The existing system (State Electricity Board structure) in some utilities may continue in the transition period with functional and managerial improvements (as a few utilities have done creditably well even in the existing dispensation of vertical integrated structures) may be allowed.

Setting up Electricity Regulatory Commissions

In order to provide a congenial and confidence building climate and to promote healthy competition and reasonable and transparent tariff making arrangements, the Central Electricity Regulatory Commission (CERC) has been set up at the central level. The CERC, has been charged with the following responsibilities :

(i) to regulate the tariff of generating companies owned or controlled by the central government;

(ii) to regulate the inter-State transmission including tariff of the transmission entities;

(iii) to regulate the inter-State bulk sale of power;

(iv) to aid and advice the central government in formulation of the tariff policy etc.

The process of unbundling of generation, transmission and distribution, establishment of regulatory mechanisms and readjustments in the existing structures are at different stages of implementation in different federal utility systems. In Orissa, the State Electricity Regulatory Commission has since been set up. Regulatory systems would be on anvil in Rajasthan and Gujarat. On the other hand, in States like Andhra Pradesh, Maharashtra and Tamil Nadu, attention is being paid for readjusting the existing structures. In other States, the choice of the unbundled systems is at different stages of discussion and debate. Other areas that would come under the purview of the commissions (at the State level) will be regulation of the working of licensing, protection of the interest of the consumers, ensuring reasonability in the operational efficiency and pricing of electricity.

Environmental Management Reforms

Environmental management and its internalisation in the project conceptualisation, operation and maintenance has also emerged as one of the key factors in the reforms process in the National Agenda. The changing economic structure, in line with the global economic and commercial environment for liberalization and deregulation, has also been given due weight in the planning for environmental upkeeping related to developmental activities.

Electricity has its own direct and indirect contribution to environmental degradation over the past decade and a half.

The following attempts have been made to :

(a) Assess the stresses on the environmental balance against the backdrop of changing economic policies and programmes, including the developmental activities;

(b) Review the current programmes in the implementation of energy development projects and address the associated environmental problems;

(c) Identify the future direction and thrust for programmes to establish priorities, and outline the strategy (environment-development related); and

(d) Identify programmes and projects for improving and provisioning of the environmental services in developmental activities, and integrate the environmental concerns in the development projects.

Above objectives are endeavoured for realization through appropriate structural instruments and strengthening them for their easy implementability. The action programmes and studies include:

(i) strengthening the area of environmental assessment,

(ii) increasing the environmental awareness through suitable demonstration programmes and facilitating their familiarization.

In respect of thermal power generation, attention is being paid to pollution abatement measures and integrating them in project conceptualisation and operation. Attention is being paid to the following concerns through appropriate technologies :

(a) Emission of pollutants into the atmosphere;

(b) Water pollution due to the discharge of effluents from power stations;

(c) Temperature changes in discharged water in ponds from the power stations;

(d) Flyash disposal effects; and

(e) Sulphur effects (though Indian coals are characterised by low sulphur content).

Environmental Impact Assessment (EIA) studies are made part of the project appraisal activity and stipulations on safeguards and measures for abatement are made part of the study which include the following :

(a) Information on dust collection and suppression equipment;

(b) Inception of a green belt around the power plant, and

(c) High efficiency ESPs and multi-fuel stack to reduce the ground concentration.

Besides EIA studies, as a part of the project appraisal, independent case studies are conducted during the period of operation so as to reiterate the concepts envisaged at the time of appraisal.

In case of hydroelectric projects, Environmental Action Plans are stipulated before the project clearance is arranged—the nature of the stipulation varies depending upon the requirement at specific site/sites. The general areas covered, however, include the following:

(a) Rehabilitation Master Plan for the project-affected people;

(b) Flora, fauna, forest and wild life protection;

(c) Catchment Area Treatment Plan (CAT);

(d) Rehabilitation and relocation of archeological, religious and historical monuments;

(e) Compensatory Afforestation Plan; and

(f) Disaster Management Plan.

Recent Legislative and Policy Measures

The following legislative and policy measures have been initiated recently :

Legislative Measures

— Electricity Laws Amendment to facilitate transmission.

— Electricity Regulatory Commissions Act enacted.

Policy Measures

— New Liquid Fuel Policy announced.

— Energy Conservation Bill to ensure optimum utilisation of energy to be finalised shortly.

— Central Electricity Regulatory Commission established.

— Revised Counter Guarantee procedure finalised.

— Techno Economic Clearance (TEC) given for 48 projects.

— 4% interest subsidy on PFC loans under accelerated generation and supply, for

Ninth Plan period, for expeditious completion of renovation, transmission and generation projects of State and Central power utilities.

— Additional 1% rebate of PFC loans to States setting up Regulatory Commissions.

— Special Action Plan and Package for North East States including additional 1% subsidy on PFC loans.

— Public investments in power sector raised to Rs. 9,950 crores in 1998-99, as against Rs. 8,188 crores in 1997-98.

— State governments empowered to clear projects facilitating increased electricity generation.

— Efficiency of existing power utilities improved, facilitating increased electricity generation.

— Comprehensive Hydro Policy to facilitate more hydro power projects being finalised.

— Introduction of Availability Based Tariff for rationalising power tariff structure and stabilising availability of power.

— Policy for facilitating Mega Projects to be announced shortly.

Oil Sector Reforms

Functional Reforms (Investment Promotion)

Reforms in the oil sector include allowance to private participation in the exploration, production and marketing activities of the crude and marketing of specific oil products in selected areas. Initiatives are taken for rationalisation of pricing policy for the individual products for greater reflection of the real costs on the prices. Joint ventures in the exploration and drilling of new wells will be encouraged. Deregulation of the natural gas sector to encourage private participation in exploration, production and transportation of the natural gas within the country is on the anvil. This may pave the way for independent regulation for gas transportation and distribution.

Deregulation in the procurement, management and use of oil products are presently envisaged in steps. Prices of oil products such as naphtha, fuel oil, diesel and LSHS will be related to the landed costs. Steps are proposed for abolition of controls on consumer prices. Nearly 80% of all petroleum products consumed in the country are likely to be linked to international market prices.

The price of natural gas will be linked to the international basket of fuel oil prices — comprising a mix of low and high sulphur fuel oils—in order to make complete parity with the international oil prices subsequently. The price of gas to the consumer has been linked with the calorific value.

Structural Reforms

In order to make the functional reforms workable and arrange a congenial atmosphere for private participation, regulatory mechanisms through a coordinated approach at the national level has been initiated with the establishment of a Directorate General of Hydrocarbons (DGH) as a monitoring entity for regulation. In the downstream, the Oil Coordination Committee (OCC) was set up to provide technical and operational support in the implementation of regulatory policies. Presently, the OCC is operating as a national downstream corporate planning-cum-supply planning entity. The role of OCC is likely to be adapted in order to make the OCC a monitoring agency.

Coal Sector Reforms

Functional Reforms (Investment Promotion)

The Coal sector reforms include offers for participation by private sector for mining in potential blocks for augmenting the annual coal production. Legislative measures to effect deregulation in the coal sector are being completed.

The deregulation includes :

(i) freedom to consumers to choose the coal supply sources;

(ii) negotiation of quantity and price in a market-driven atmosphere and competitive climate;

(iii) measures to diversify the choice for the consumer in terms of resilience in the indigenous supply; and

(iv) import of coal.

Competitive bidding for offering new mine development programmes in the private sector would be introduced. Setting up of the Regulatory Commissions and adaptations in the institutional structure, encouragement of joint ventures between the existing public sector core companies and private sector will also be encouraged.

In order to make the coal industry financially viable and self-supporting, the following measures have been initiated :

(i) Rationalisation of the manpower.

(ii) Decontrol of prices of all non-coking superior grade coals and coking coals.

(iii) Closure of uneconomical mines.

(iv) Capital restructuring packages for collieries.

(v) Reduction of import duty on equipment and spares.

(vi) Parity for distribution of imported coal and indigenous coal.

(vii) Implementation of cash and carry system for coal supplies.

Structural Reforms

The structural reforms aimed at encouraging the private participation in coal mining and build up of washeries etc. include the following :

(i) Provision for captive coal mining by consumers.

(ii) Construction of washeries on Build-Own-Operate (BOO) basis.

(iii) Setting up of washeries by power stations.

(iv) Setting up of washeries by private agencies as agents to the power stations.

Environmental Management Reforms

The annual production of coal is about 300 mt. About 200 mt of low grade high ash variety of coal is supplied to power stations. This results in high level of ash and high emissions. In line with the decisions of the Framework Convention on Climatic Change (FCCC), the emissions have to be controlled in order to contain the greenhouse gases (GHG). In the above context, from 2001 AD, beneficiation of coal has been made mandatory in order that the ash content does not exceed 34% for all power stations which are located at a distance beyond 1000 kms from the coal supply source. Use of beneficiated coal has also been made mandatory in non pit-head power plants which are located in the urban centres/sensitive or polluting areas.

The Coal Mines Nationalisation Act has been amended in order to permit coal washing in the private sector. Coal rejects from washeries containing adequate heat value are proposed to be used for power generation in specifically designed boilers. Mandatory use of ash for different commercial and service purposes is likely to be insisted upon in the near future to manage large ash production within permissible limits.

ENERGY TECHNOLOGY PERSPECTIVES

Overview

The strategy for technology absorption, adaptation and indigenisation have the following basic elements :

— Development and indigenisation of new techniques of exploration for augmenting the energy resource base, in both commercial and non-commercial forms;

— Development of technologies to maximise energy production and recovery;

— Improvements and adaptations of the existing technologies with emphasis on conservation;

— Development of new, efficient and mature technologies for utilising indigenous resource of poor quality;

— Development of technologies for harnessing alternative sources of energy.

The Technology Policy of the government aims at reducing the vulnerability of the conventional energy technologies and their capital intensiveness. It aims at striking a balance in the four basic elements of supply, consumption, costs and environment, in order to make the energy economy financially manageable and environmentally sustainable. In striking such a balance and prioritisation, adequate attention is being paid to focus on short term gains without impairing the long term interests.

Specific technology perspectives in sub-sectors are as follows :

Power (Conventional, Thermal and Hydro)

Technology upgradation in the electricity including the nuclear power generation has covered the entire gamut from generation to distribution to the ultimate consumer and characterising the cost and conservation of electrical energy in economic and environmental terms. Technologies for renovation and modernisation and O&M were developed in the 80's.

Following areas for thermal power technologies have been stressed upon :

— Reheat Units and use of improved materials and designs;

— Design margins, redundancy and preventive maintenance;

— Fluidised Bed Combustion (FBC) technology for minimising environmental impact.

— Coal Combustion during the pre-combustion and combustion stages.

— Bubbling Fluidised Bed Combustion (BFBC) technology in 30-60 MW range.

— Circulating Fluidised Bed Combustion (CFBC) technology for larger capacity units.

— Integrated Gasification Combined Cycle (IGCC) involving conversion of coal.

— Fly ash based building material production units near ash disposal points.

On the hydroelectric generation side, the areas of interest for technology upgradation are :

— suitable civil engineering measures as well as adoption of new materials like ARCLOCK, BOFORS and coatings such as plasma nitriding for runners and other water borne components for tackling silt related problems after field trials;

— technology for site-assembly of split transformers and runners for overcoming transport limitation;

— Adjustable Speed Generating Units for pumped storage and conventional hydro plants.

— Digital Turbine Governor to enhance performance of the generating units.

Developments in a number of areas such as welded shafts for high capacity turbines, self lubricated type thrust bearings, polymer trash racks, ultrasonic devices, butterfly valves etc would also have to be initiated.

In the field of transmission and distribution, following areas merit consideration :

— higher use of power electronics;

— use of multi-circuit, multi-voltage lines;

— adoption of new designs for towers like narrow-based, guy tower, etc;

— uprating and upgrading of existing transmission lines;

— use of underground power cables;

— introduction of SF6 breakers; and

— adequate provision of better quality meters.

The organisations/institutions responsible for the technology upgradation activities in the power sector include :

— Central Electricity Authority

— State Electricity Boards (in federal States)

— National Thermal Power Corporation

— National Hydroelectric Power Corporation

— North Eastern Electric Power Corporation

— Central Power Research Institute

— Bharat Heavy Electricals Limited

— Central Board of Irrigation and Power

Oil and Natural Gas

There has been considerable step-up in the technology upgradation in the oil sector (exploration and production), particularly over the past two decades. Improvements in technologies for exploration, also led to profitability for development of marginal fields; reduction in the exploration risks; reduction in costs; and above all, contribution to broadening the meagre resource base. In addition, adoption of new techniques/technologies contributed to harnessing the associated flared gas economically in the production process.

Technology focus in the oil and gas sector include the following :

— Improving the oil exploration techniques.

— Development of enhanced recovery techniques.

— Modernisation of drilling technologies to harness energy from deep wells.

— Modernisation of oil refining techniques and technologies.

— Development of products with varied quality in the refining processes.

— Development of technology for exploration and exploitation of Coal Bed Methane.

— Development of transportation technologies for LNG and CNG.

There are several organisations, each specifically concentrating on technology promotion in the oil sector. These include the following :

— Institute of Petroleum Exploration

— Institute of Reserve Oil Studies

— Institute of Drilling Technologies (exploration and oil well development and production).

— Indian Institute of Petroleum (refining and refinery development)

— Engineers India Limited (conceptual design and engineering aspect of the oil production and management)

— Indian Oil Corporation (marketing and conservation of oil products)

— Regional Research Laboratories (under the Council of Scientific and Industrial Research)

— Petroleum Conservation Research Association (PCRA) (conservation of petroleum products and environmental management)

— Oil Coordination Committee (technology policies and packages)

Coal

In the coal sector, technology focus/adaptation of techniques include the following :

— Development of new and modernisation of existing techniques in coal exploration.

— Investigation of specific mining problems.

— Modernisation of technologies in coal mining to achieve improved coal recovery.

— Reduction of cost and improving mine safety.

— Beneficiation technologies for environmental benignness.

— Gasification and liquifaction.

— Substitution of petroleum by coal.

— Technology for harnessing new coal products — Coal Bed Methane (CBM)

— Improved coal combustion techniques like Fluidised Bed Combustion (FBC) to augment utilisation of low quality coal.

The major research institutions working for the above and other objectives in coal technology include the following :

— Central Mine Planning and Design Institute (CMPDI).

— Central Mining Research Institute (CMRI).

— Central Fuel Research Institute.

— Regional Research Laboratories. (subsidiaries of Council of Scientific and Industrial Research, Delhi)

— Bharat Heavy Electricals Limited.

Nuclear Power

The main thrust in the nuclear power programme have been to continue with the natural uranium fuel in the HWR power plants and pursue the R&D and technology activities relating to development of FBR technology in order to enable utilisation of the much larger reserves of nuclear resources available in the country. The thrust, therefore, includes:

— stabilisation of HWR technology

— development of FBR

— development of nuclear fuels (including reprocessing of fuels)

— stabilisation in the nuclear safety and health

— stabilisation in the nuclear safety management

— upgradation of unit sizes.

The institutions involved in the technology upgradation of nuclear fuels are :

— Bhabha Atomic Research Centre at Trombay.

— Reactor Research Centre at Kalpakkam concerned with reactor technologies.

— The Atomic Minerals Division relating to prospecting and exploration of atomic minerals.

— Tata Institute of Fundamental Research and Saha Institute of Nuclear Physics (R&D support in nuclear area).

Non-conventional Energy Sources

Development of non-conventional source of energy particularly in the areas of solar (photovoltaic and solar thermal), wind, biogas technologies are given technology thrust and are already gradually approaching commercial comparability for costs with the conventional power generation technologies. Opportunities are being created for R&D programmes for power generation from woody biomass, agricultural residues etc. A major programme of bagasse based cogeneration aiming at optimising generation of surplus power and supply the same to the grid is a priority. Through various initiatives, grid quality power generation from renewables are gradually attaining maturity.

The Institutions involved in the development of new and renewable sources energy are :

— Solar Energy Centre;

— Centres for Development of Non-conventional Energy Sources in the federal States;

— Indian Institutes of Technology;

— Indian Institute of Science;

— National Laboratories under the Council of Scientific and Industrial Research.

ENERGY PERSPECTIVE 2020

Issues in Perspective

The population of the country may exceed 1.3 billion by the end of 2019-20 involving urban/rural population ratio at 43%/57% in that year. With rapid urbanisation and possibilities of literacy improvement in the rural areas, followed by high rates of economic growth in the range of 6-7% is likely to be accompanied by appreciable levels of increase in the per capita income and hence the life styles. This will have obviously great influence on the energy demand as well. There is a growing perception that environment protection and conservation would have to be adequately internalised in the growth of the energy sector. The key issues involving the energy demand implications therefore include :

(i) rise in population;

(ii) sustainability of economic growth;

(iii) access to adequate commercial energy supply;

(iv) accessibility to financial resources needed to support expansion of the energy supply systems, and

(v) rational pricing regime.

Energy Demand 2020

Estimation of demands for energy sector as a whole in a country of India's size and diversity is nebulous. Customarily, demand projections for individual sub-sectors of energy consumption viz. coal, oil, electricity etc. are made at regular intervals with adequate effort for analysing the impacts of the economic policies on the energy use in the sub-sectors. A few studies were also made intermittently by agencies in the nature of analysing the energy sector scenario in an integrated framework. These include the demand estimates made by the Working Group on Energy Policy in the year 1979. Emphasis in these studies was more or less on efficiency improvement in the use of individual sub-sectors, inter-fuel substitution possibilities, and sometimes involving cost optimisation. However, integrated energy demand studies involving the implications of sustainability of demands, demand management also involving the financial and environmental sustainability etc. have been limited, at the national level.

A few model studies are made by experts in the recent past for projecting the energy demands in the global perspective for the period upto 2020. The latest Study by E. A. S. Sarma, J. N. Maggo and A. S. Sachdeva indicates the overall energy demand perspectives for three different scenarios as follows :

(i) a referenceable scenario in which the business as usual, (BAU) would be allowed to continue,

(ii) an efficiency oriented scenario (EFF) considering the scope for reduction in the energy intensity based on induced efficiency through technology and management adaptations, and

(iii) an environmentally constrained scenario (ENV) considering the implications for emphasis on minimisation of adverse environmental impacts in augmenting the energy supply system.

Projections were made under the three different scenarios for the different sectors of consumption and for three terminal years viz. 1999, 2009 and 2019. Projections made under the above for the "business as usual" scenario indicate the total demand requirement by 2019 would work out to 39107 peta joules (934.0 mtoe) as compared to 9727 peta joules (232.3 mtoe) in 1989. However, looking from the point of view of environmental sustainability and huge investment involvement, energy conservation and energy efficiency programmes would have to be focussed. Their successful implementation can reduce the demands substantially.

In order to meet the level of final energy demand projected in the BAU Scenario, the installed capacity in power generation will need to increase from the present level of 1,00,000 MW (comprising 86,000 MW in utilities and 14,000 MW in non-utilities) to nearly 3,85,000 MW in the year 2020. The coal and oil requirement will correspondingly be 1,400 mt and 400 mt respectively.

ROLE OF NTPC IN INDIA'S POWER SECTOR

7

Rajendra Singh
Chairman and Managing Director
National Thermal Power Corporation Ltd.,
New Delhi (India)

The rare feat of an organisation exceeding the expectations of the major stakeholders, that too amidst limitations and through crises, is a business watcher's delight. That NTPC has achieved this rare feat is expressed in the words of no other institution than the World Bank itself which is not known to be too generous in its praise. The World Bank has observed

> "......... Yet, despite the odds, NTPC has been an all-too-rare institutional development success, achieving virtually all that was expected of it and, in some areas, more"

A study conducted on the basis of 1996 data by the UK based Marketline International, ranked NTPC as the world's third most efficient generating company and the ninth highest electricity generating company in the world. The identification of NTPC as a potential global giant is a natural corollary of the Company's superlative performance and rich potential.

Seeking to build upon its strengths and to capitalise upon the possible opportunities NTPC, in today's globalising and competitive scenario, has given to itself a new Corporate Vision which sums up the role it has assigned to itself in the Indian power sector and in the Indian economy. The Vision statement says:

> "NTPC, a front runner in the Indian power sector, to be one of the largest and best power utilities of the world and thereby contributing to India's emergence as one of the world's leading economies"

In order to translate its new vision into reality, NTPC is finalising a long term Corporate Plan for the period upto 2012 A.D.

With all-round robustness, NTPC is today generating one-fourth of the country's power with one-fifth of its total installed capacity. The performance of NTPC is best summed up in the following words of the World Bank Performance Audit Report, 1993:

> "NTPC has demonstrated that Government owned power utilities can be operated at efficiency levels comparable to those of privately owned utilities in India and well run utilities outside of India"

The confidence of NTPC stems from the robustness of its past performance. The exciting and achievement filled journey of NTPC began in the year 1975.

The Growth and Accomplishments of NTPC

Beginning its journey in 1975, NTPC has emerged as the largest power utility in the country (with an installed capacity of 16,795 MW and approved capacity of 20,515 MW) and as the largest Company in the country in terms of approved investment (over Rs. 36,000 crore).

Since commissioning of its first generating unit in 1982 at Singrauli, NTPC has gone on setting new standards in the area of project implementation.

Thus, in a period characterised by time and cost overruns by most PSUs and by 'sickening' of units by many private entrepreneurs, NTPC came to receive accolade from several quarters for effective project management.

Initially NTPC specialised in setting up large pit-head coal based stations and EHV transmission lines. Subsequently, the Company also brought the combined cycle technology in gas based power stations where it has gone for multi-fuelling (e.g. Naphtha) option to meet the exigencies arising out of fuel shortages.

NTPC also set up high voltage transmission lines. This business has been transferred to the Power Grid Corporation of India Ltd., but the strength acquired in this area is being used in serving several international consultancy assignments. NTPC introduced the High Voltage Direct Current (HVDC) technology of power transmission in India.

Backed by a sound Human Resource Management and a team of over 24,000 excellent power professional and having a sound and change-sensitive financial/commercial strategy, NTPC has been doing very well on the various performance parameters.

A brief recap of the Company's performance during the Eighth Five Year Plan spanning the fiscals 1992-93 to 1996-97 as also the performance during the fiscal 1997-98 demonstrates the all-round excellence of NTPC:

Operational Performance

Improvement of PLF and operation efficiency have always received sustained thrust in NTPC. For Regions other than the Eastern Region, a PLF of 84.5% was achieved during 1996-97. During the year 1997-98, too, the overall PLF would have been much higher than 75.2% but for the poor 45.7% PLF of the Eastern Region caused due to grid restrictions and evacuation constraints. Specific strategies like ensuring adequate fuel supply, shifting focus to preventive maintenance, focusing on major recurrent causes for forced outages etc. have helped achieving the desired results, as may be seen from the table given below:

	Unit	*91-92*	*92-93*	*93-94*	*94-95*	*95-96*	*96-97*	*97-98*
Generation	MU	61382	66113	76478	79091	93155	97609	106288
PLF	%	70.59	70.00	78.07	76.57	78.80	77.00	75.2
Availability factor	%	83.50	83.34	86.48	85.92	85.32	84.1	85.03

Productivity

In the past six years, the employee productivity has almost quadrupled as measured by Turnover/employee. Also it is worth noting that for an increase of 48.2% in installed capacity, the employee strength went up by only 13.5%. This, inspite of absorbing 2124 number of employees alongwith the 460 MW Talcher Thermal Power Station from OSEB which had a large number of surplus employees.

	Unit	*91-92*	*97-98*	*% increase*
Installed capacity	MW	11,333	16795	48.2
Generation	MUs	61,382	97609	59.0
No. of employees	No.	21,498	24394	13.5
Generation/employee	MUs	2.86	4.50	57.3%
Turnover/employee	Rs. lakh	17.93	52.63	193.5

Capacity Addition

During the Eighth Plan period, NTPC added 5002 MW by completing all the on-going schemes and 460 MW by taking over Talcher TPS, thus taking the installed capacity to 16,795 MW. NTPC invested over Rs. 10,700 crore in plan period with almost no 'net budgetary support'. In fact, NTPC added 31.5% of the total capacity added in the country during the Eighth Plan.

NTPC is planning to add over 14,000 MW to become a 30000+ MW Company by the year 2007 and more projects are being examined to be taken up during IX and X Plans. During the IX Plan, the Company plans to add 6300 MW and for the X Plan period, it has already identified projects aggregating over 8000 MW.

Financial Performance

Financially NTPC has done very well in the past six yeaRs. Turnover has more than trebled from Rs. 3419.41 crore in 1991-92 to Rs. 12,620 crore in 1997-98. The profit soared from Rs. 734.62 crore (excluding transmission lines) in 1991-92 to Rs. 2122.30 crore in 1997-98. The dividend rose by almost eight times from the maiden amount of Rs. 65 crore in 1993-94 to Rs. 503 crore in 1997-98. The detailed financial performance is shown below:

	91-92	*92-93*	*93-94*	*94-95*	*95-96*	*96-97*	*97-98*
Turnover	3419	4485	5918	6486	8545	10394	12620
Return on Capital Emp.	10.72	10.90	11.51	11.41	11.79	11.91	12.03
Return on Net Worth	1.95	11.57	11.73	11.90	12.15	12.15	12.33
Net Profit	734.62	886.57	1057.97	1125.00	1352.61	1679.43	2122.30
Dividend	—	—	65	130	195	405	503

MOU Target Achievements

NTPC is among the first PSUs which entered into the Memorandum of Understanding with the Government in 1987-88 and has since been signing an MOU with the Government every year. NTPC has maintained its performance under the 'Excellent' category (the best category) ever since the MOU system became operative.

Single Largest Beneficiary of World Bank Funding

As a result of its impressive track record in financial management, the World Bank and various other multilateral funding agencies like OECF, ADB, KfW etc. have reposed immense faith in NTPC by extending financial support for its growth. Consequently, today NTPC is the single largest beneficiary of the World Bank funding (with an assistance of about US $ 4 billions) in the country.

Commitment to Quality

NTPC stands consistently and comprehensively committed to quality in all aspects of its functioning. It has been awarded the ISO 9001 Certification by the internationally reputed inspection agency M/s Lloyd's Register Quality Assurance Ltd. (LRQA), under accreditation from National Accreditation Council of Certification Bodies (NACCB), U.K. and TGA-DAR, Germany.

NTPC has its job cut out. It is poised to take on the challenges which the Indian power scenario offers in ample measure. The Indian power scenario is too complex to allow for a simple description and the sectoral issues are too varied to allow any simple prescription.

The Indian Power Scenario

Fifty years down the road since Independence we are among those at the rock bottom in terms of per capita energy consumption although we have made substantial progress by developing over 88,000 MW power capacity compared to a meagre 1360 MW in the year 1947. Although, in absolute capacity terms, we are comparable to U.K., Germany and France, in per capita terms, we are just nowhere near them. Even China, which began its development around the same time as we did, is far ahead of us in both absolute (over 2.25 lakh MW) and per capita terms. This is not to suggest that we have not achieved much. We have had certain sterling achievements, many of them aided by the power sector (agricultural growth, for example) and whatever we have done has been accomplished under a pluralist democracy which, at times, takes longer to arrive at consensus or to implement decisions compared to the system which obtains in a one party-run State like China.

But all things considered, we are capable of achieving much more and at much faster speed than what we have achieved so far. Completing fifty years since Independence and preparing to welcome the new millennium, we must kick-start our infrastructure sector in general and power sector in particular so as to be able to find our rightful place among the leading economies in the world.

To meet the 'power challenge' effectively under our unique circumstances, we need to maximise our power supplies at minimum possible investments. Optimum utilisation of our existing capacity provides a cost effective and prompt way of creating substantial supply of power. Augmenting our existing capacity through Renovation and Modernisation of existing plants, curbing Transmission and Distribution losses and improving the performance and reliability of the plants, we shall be able to supply substantially greater amount of power from the existing capacity at much less costs compared to those required for adding new projects. But we have to add greenfield projects too. All this has to be done alongwith restructuring or realigning of the major players in the power sector like the State Electricity Boards, the RLDCs, the IPPs, CEA etc. Restructuring of the power sector, reform of the State Electricity Boards (SEBs), rationalisation of tariff, improvement in not only generation—but also in distribution and transmission and above all—a commercial re-orientation of the entire sector are necessary for toning up the performance of the sector. Then only shall we have enough finances flowing in on a reliable basis so as to meet our capital requirements for capacity addition.

The immediate challenge before the Indian power sector is to narrow down the gap between the demand and supply. The energy shortage is approaching 10-11% and the peak-demand shortfall is in the range of 20%. This has bottlenecked our economic growth. According to an estimate, the opportunity cost of the energy shortage and the peak-demand shortfall is 1.5% of the GDP. We need over 55000 MW of new capacity during the IX Plan. But the feasible addition may be only in the range of 40000 MW. Even this has a question mark against it. Even this 'feasible' addition alongwith the associated transmission and distribution network calls for massive financial resources amounting to well over Rs. 200,000 crore (US $ 47 billion). If we take into account the Tenth Plan (for which the investment planning should ideally begin right now), the financial resources required are that much more.

Financial health of SEBs holds the key to the much needed investor comfort and capital flow into the power sector. Restructuring SEBs along commercial lines may help attract investments. This needs to be done, and quickly. We must capitalise on the fact that most of the investors are still interested in India as the largest emerging power market with a sound legal system, its political-financial challenges notwithstanding. Alongwith China,

Brazil and Turkey, India promises to constitute the power-quartet in terms of international power growth. Also, the South East Asian crises have stalled many projects in that region which were vying for finance.

While we need a rich infusion of foreign investment, we will have to attract domestic investors in a big way. The Rakesh Mohan Committee's Infrastructure Report also says that the lion's share of investment in the infrastructure sector, including power, has to come from the domestic market. To make investors interested in the power sector a wide range of steps beginning with power sector reforms have to be taken. The SEBs need to be restructured. Any of the models suggested by the Shankar Committee or any variant thereof may be adopted by the SEBs. Reforming and ultimately "privatising" distribution and transmission shall infuse money and vibrancy. The sensitive yet essential step of rationalising tariff to make SEBs viable and credit worthy will have to be taken. Special tax incentives may be extended to small investors subscribing to 'power bonds' etc. Faster clearance of projects should also be ensured. The idea is to make investment attractive and the power projects viable and bankable. The frequent policy shifts tend to increase the perceived risk and therefore the risk premium. Hence we need a stabilised and enabling environment conducive to investments.

There is another workable means of mitigating, to some extent, the complex problem of financing some of the activities in India's power sector. A nominal cess of 10 Paise/KWh can be levied on energy sold in the country. With this, a Central Fund called '"Power Development Fund' with an estimated Rs. 20,000 crore in the Ninth Plan and about Rs. 45,000 crore in the Tenth Plan could be raised. This could finance our Renovation and Modernisation (R&M) as well as Transmission and Distribution (T&D) improvement efforts and could also partly meet our new capacity investment needs. China, whose power development have seen enormous growth had successfully tried this method by imposing power development tariff at the rate of 2 cents per unit and the realisations have helped China's capacity addition programme.

The issue of fuel and appropriate power-mix is also of crucial significance. The thermalhydel mix is disproportionately lopsided with hydro having only 24% of the total installed capacity. Since India has a large hydro potential (130 GW), the hydro option should be given greater thrust. In spite of the long gestation period and large scale displacements involved, many experts are of the opinion that developing mega hydel projects is a prudent policy for the long term power planning in India. The Small Hydro Power (SHP) should also be given equally greater thrust. The Eighth plan target for SHP was 600 MW, but the actual capacity installed was less than 60 MW. A hydro power plant (micro, mini, medium, or mega) is a viable option due to its pollution free generation utility for managing peak load, and abundant scope in India. The aspect of resettlement of the affected population can be better planned and implemented to obviate the problems on this score. NTPC has decided to enter into Hydro power generation and is identifying projects. This will help increase hydro generation in the country.

As for gas, it has limited availability. While several schemes are being talked about to import LNG, the lead time and costs alongwith other implications are major variables which will determine its viability. The unique conditions of the Indian fuel market have made naphtha a popular fuel. However, naphtha too is not a prefered fuel for power generation, for reasons of cost and inherent risks of handling. LNG on the other hand is a preferred eco-friendly option with adequate availability, flexibility of sourcing and safety in handling. However, a judicious decision would be needed, to the extent of LNG import in view of the foreign exchange risk. Currently, 25% of the world's gas is being used in the form of LNG and it is going to rise further. An agency dedicated exclusively to LNG can be more effective in providing LNG to India's power sector. West, South, and South-East Asia including Bangladesh, have good reserves of natural gas. There is an urgent need to explore oil fields in India for natural gas which may provide the answer.

All things considered, coal is going to be the mainstay of India's power generation in the short and medium terms. With appropriate technological breakthroughs, it may even remain a long-term favourite. About 70-80% of India's power capacity is to be coal fuelled. But we need intensive mining and adequate transportation to be able to meet the coal demand of about 350 Million Tonnes per annum during the next few years as against our supply of 200 Million Tonnes. Upgrading and expanding port facilities and surface transport remain vital to improve coal and fuel availability. In fact, India is in urgent need of short, medium and long term fuel policy and a programme for optimum energy mix.

Renovation and Modernisation is of crucial significance to improving India's power situation. Although the average national PLF of thermal plants rose significantly from just over fifty percent to nearly sixty-five per cent over the last decade, there is much scope for further improvement. According to an estimate, about 15,000 MW thermal capacity is generating at below 45 percent and about 8000 MW of hydel plants are performing well below the rated capacity. R&M of the thermal plants alone can quickly result in an output increase equalling 5000 MW of new capacity. In fact, in the longer term, the R&M initiatives will yield much bigger returns with much lower costs (only 25 to 40 percent of those required for adding new capacity). Hence, there is a need to give much sharper focus on R&M and infuse much greater investment (compared to the paltry two to three percent as was being done) and involve the private sector and others in these efforts. It has been seen that R&M has resulted in 20 to 60 percent improvement in Indian plants. In this context, the turn around stories of a few NTPC projects are worth mentioning.

Turn-around Expertise of NTPC

One remarkable aspect of the achievements of NTPC is the turning around of under-performing units of SEBs. The Feroze Gandhi Unchahar Power Station was acquired by NTPC as part of a win-win deal with the Uttar Pradesh Government whereby the dues of UPSEB were adjusted and the expertise of NTPC turned around the new yet languishing power unit. The remarkable speed and dimensions of the turnaround can be seen in the table below:

Unchahar [Taken over in February '92]

Parameter	*Prior to takeover*	*12 months after take-over*	*1997-988*
Availability (%)	27.22	78.96	90.8
PLF (%)	18.02	73.69	80.1
Sp. Oil (Ml/Kwh)	21.83	3.33	1.03

An even more challenging turnaround story is being scripted by NTPC at the OSEB's old and nearly rickety power plant at Talcher taken over in 1995. The table below indicates the dramatic gains in such an old power plant. While NTPC has already bettered the PPA commitments, it is poised to, turn-around the plant in a much shorter time frame than is usually taken.

Talcher [Taken over in June '95]

Parameter	*June 95*	*1995-96*	*1997-98*
Availability (%)	37.28	54.45	77.0
PLF (%)	18.7	29.52	52.0
Sp.Oil Conspn. (Ml/Kwh)	22.57	18.69	4.01

From the view point of capital scarcity in our country, turning around such old units is a low cost, but high and quick return option. Thus, these experiments helped NTPC, the concerned SEBs and the entire nation in terms of economy as well as power availability. NTPC has also successfully turned around the Badarpur Thermal Power Station being managed by it, as can be seen from the following table:

Badarpur TPS [Taken over in April 1978]

Parameter	*1977-78*	*1996-97*	*1997-98*
Availability factor :	40.0	83.16	90.2
PLF (%)	31.94	66.14	72.5
Sp. oil conspn. (Ml/Kwh)	15.55	2.22	4.64

In fact, encouraged by its experience and considering the overall sectoral advantage through R&M, NTPC has embarked on the R&M path in a big way. It is in the process of identifying a Joint Venture partner for launching an R&M Joint Venture which will exclusively look after R&M of old SEB plants.

Some More 'Thrust Areas' for NTPC

NTPC will attempt to maximise the exploitation of existing capacities at the most economical costs. The R&M intervention in the projects taken over by it have already proved the efficacy of this approach. Now NTPC is going to set up joint venture enterprises with leading power-players to take up R&M jobs of old and ageing SEB units.

Efficiency improvement and environment management are significant issues calling for concerted response. In this context, too, NTPC has taken steps worth emulating by others in the sector. Centre for Power Efficiency and Environmental Protection (CENPEEP) set up in 1994 by NTPC, with the assistance of USAID and the achieve involvement of the US Department of Energy, is a national resource centre. NTPC, the largest power utility in India, has become the agency to implement the Greenhouse Gas Pollution Prevention Project (GEP) of the US Government in India's power sector through CENPEEP. The Centre is involved in the task of achieving improved efficiency of coal-based thermal power plants through demonstration and dissemination of new technologies and practices related to efficient coal conversion better plant availability and reliability. Increased efficiency helps reduce $C0_2$ emissions. Commercialisation of ash-utilisation technologies and emissions monitoring, etc. are also done by the Centre with good results. The project aims at reduction in emission of greenhouse gases per unit of electricity generated by increasing energy productivity. Typically 10% of capital expenditure of each NTPC plant has gone into environmental equipment and measures. NTPC has collaborated with E-7, a group of eight of the world's leading utilities in the West, towards efficient and environment friendly power generation.

Ash utilisation is given adequate attention. NTPC utilises 10-11% of ash generated against the national average of 2-3%. And NTPC will make sustained efforts to encourage ash based industries. And NTPC will make sustained efforts to encourage ash based industries.

In this context, India should gradually adopt super critical cycle parameters for large units. It should also adopt IGCC and other state-of-the-art technologies which will make it predominantly coal based power development environmentally more compatible, besides resulting in high output and productivity, and NTPC will take steps to adopt such technologies.

Also adequate measures must be taken to effectively resettle and rehabilitate the people affected by power project. We are developing power for the people and not at their cost. Our power development must always have a human face. NTPC has adopted a comprehensive Resettlement and Rehabilitation policy for the Project Affected Persons and will promote such initiatives in the sector.

Promoting skills is very necessary and 'knowledge' is going to be the key determinant of corporate competitiveness. NTPC has set up an extensive training network at the apex of which is its Power Management Institute (PMI) which is committed to adopting innovative ideas in the field of power management. PMI is a facility open to the entire power sector. And some of the leading private organisations (in addition to SEBs) like ABB, ESSAR etc. have utilized PMIs facility for training their personnel.

A lot needs to be done before we are actually 'power-driven' into becoming a major economy. While the industrialised world is moving towards a situation where 'brain-power' industries and service-sector are going to have ascendancy and the entire world is inexorably moving towards being knit as a global village, we must take giant strides to join the league of the leading economies. In fact we almost have to leapfrog to join them. For this, we need to give an urgent 'quantum-push' to the power sector.

We in NTPC are concerned about the drop in the GDP growth to 5 per cent in 1997-98 from the high of 7.5 per cent in the last three years of the Eighth Plan (1994-95 to 1996-97). At a fundamental level, the growth slow-down is attributable to the less than adequate quality and quantity of the basic infrastructure services such as power, railways and roads. The 'power-slowdown' was caused by a combination of factors. But NTPC perceives the future scenario as one full of challenges and opportunities where massive capacity addition and revamping will be done and where efficient state enterprises like NTPC will have a rightful role. We are equipped and committed to play our role in India's emergence as a major economy.

NEW APPROACHES IN POWER TRANSMISSION

8

R.P. Singh
Chairman and Managing Director,
Power Grid Corporation of India Ltd., New Delhi.

Today it is an event of high esteem for me that I am here to deliver this keynote address to the most august World Energy Council audience.

THE INDIAN POWER SECTOR

"Electricity" is a concurrent subject in India i.e. it is the responsibility of both Central government and State governments. Indian power sector has developed on regional concept of power planning and the country has been demarcated into five regions viz. Northern, Eastern, North Eastern, Southern and Western regions. Central Electricity Authority (CEA) is responsible for overall planning and coordination of power development at national level. State Electricity Boards (SEBs) are responsible for generation, transmission and distribution of power at state level. Regional Electricity Boards (REBs) are responsible for coordinating the integrated operation of the Regional grid system, setting tariffs for inter-agency power transfers and global accounting of these transfers and coordination of plant maintenance schedules.

Further, there are a number of government owned organisations with specific responsibilities and functions for utilising diverse resources in the country viz. coal, lignite, hydro, gas, nuclear etc. for power generation. In addition there are some private power generating and distribution companies to cater to local demand. It is envisaged that a total capacity addition of about 1,40,000 MW shall take place during 1997 to 2012 covering 9th, 10th and 11th Plan periods (as per the National Power Plan). The additions expected during the current Ninth plan is 40,000 MW with 45,000 MW and 55,000 MW being added subsequently during the Tenth and Eleventh Plan. However, due to resource constraints faced by Central Power Sector Undertakings (CPSUs) as well as SEBs and sanctions imposed by various multilateral agency like World Bank, ADB, OECF etc. the actual capacity addition may affect the planned capacity addition during Ninth Plan leading to further deterioration of the Power supply situation.

The investment in transmission and distribution sector till end of 8th Plan on an average has been 25 to 28% except the 1st plan when it was 53%. This has resulted into serious deficiencies in the transmission and distribution system as against about 50 percent of the total allocation of the power sector and today T&D losses are hovering at 22% on National basis although in few States the losses are as high as 50%.

Keeping in view the realities of the Centre and State Governments and State Electricity Boards (SEBs) that it would be impossible to meet the anticipated demand even quantitatively let along qualitatively, Government of India initiated the process of institutional restructuring in the year 1989 by consolidating the transmission system of various central generating organisations under single nodal agency—POWERGRID, to plan and coordinate between various generating and distribution companies as well as to act as an "honest bulk transfer facilitator for power". In addition a number of concessions have been offered to private sector investors to attract them to invest in power sector.

PRESENT OPERATIONAL PROBLEMS WITH REGIONAL GRIDS

Regional Imbalances

The Regional pattern of power development had led to imbalances in the hydro-thermal mix. As against the desired hydro-thermal mix 40:60 to private peaking support and to prevent sub-optimal level operation of thermal plants during off-peak hours, the ratio today is 25:75 on All India basis. However, hydro resources which are unevenly distributed in various regions could not be optimally harnessed due to resource constraints, gestation period of project and other bottlenecks.

For example, the hydro resources are mainly concentrated in North Eastern Region and Northern Region which approximately 73% of total coal reserves lie in the states of Bihar, Orissa and West Bengal in Eastern region. At the same time the Western and Southern region do not have enough energy resources to meet their growing demand and are dependant upon the power import from other regions. In Eastern region the demand has not developed as envisaged while planning the generation addition resulting in availability of surplus energy and high frequency and voltage levels. Efforts are being made to address these shortcomings by formulating and implementing system improvement schemes and to increase inter regional transfer of power through the existing links.

System Behaviours and Operational Discipline

The regional grids suffer from very serious problems such as wide frequency and voltage fluctuations, frequent grid disturbances, lack of grid discipline, uneconomic operation etc. In our system, load changes dictate the system frequency and generators are unable to control frequency because of low reserves and energy shortages forcing utilities to serve as many consumers as possible at a given time. Voltage fluctuations take place not only on seasonal basis but also on daily basis. During light load conditions, the voltage level is of the order of 420-430 KV whereas in the peak load conditions the voltage dips to as low as 320 KV on 400 KV buses, causing serious grid disturbances.

The reasons for this type of system behaviour is attributable to inadequate reactive compensation at all voltage levels especially at sub-transmission and distribution network. The lack of proper agreements spelling out the reliability criteria, obligations of suppliers and beneficiaries are other deficiencies in the system for operational discipline. Generating companies are generally reluctant to back down even during off-peak time as the revenue of the utilities and their employees generation incentive is linked to PLF. Similarly an SEB will not buy power from the other efficient (lower marginal cost) generating stations as the purchasing price as per present tariff arrangement is higher than the marginal cost of its own generation. There are no means of enforcing operational discipline. Into the case of an SEB drawing more than its allotted share of power, the Regional Load Despatch Centre (RLDC) can at the most request the erring SEB to regulate its demand and in most of the cases, the response from the SEB is only partial. Most generating units operate on constant load mode with turbine governors made virtually inoperative. As such, neither they participate in frequency control nor do they respond to the load fluctuations.

PIVOTAL ROLE OF POWERGRID THROUGH TECHNOLOGICAL EXCELLENCE

POWERGRID is playing a vital role to bridge the widening gap between demand for energy and dwindling supplies in this endeavour and making most economical use of limited amount of resources available through excellence in various facets including :

— Engineering

— Project Management

— Contracts Management

— Construction Management

— Operation and Maintenance

— Business Development and Consultancy

— Human Resources

Engineering

POWERGRID's sterling performance can be greatly attributed to its dedication to Technological Excellence. To cater to diverse geographical and climatic conditions of the Indian nation. POWERGRID has designed transmission towers for different geographical conditions like wind and ice, marshy, water logged and mid-stream locations besides special towers for rivers and power line crossings. Engineering expertise includes implementation of and highly specialized areas of 800 KV AC systems, 500 KV HVDC bi-pole and back-to-back station, 400 KV AC transmission system and Gas Insulated Substations.

POWERGRID is contemplating to induct state-of-the-art technology in power transmission which includes series compensation and static var compensation, flexible A.C. transmission systems, upgrading the existing transmission systems, Time Synchronization at sub-station upto an accuracy of a few micro-seconds and much more.

Emergency Restoration Systems have been acquired for faster restoration of damaged transmission lines. ERS has been deployed in Bairasaiul-Pong line, Tanakpur-Bareilly line in Northern Region and Maithon-Jamshedpur 400 KV line in Eastern Region and has saved time of 1-4 weeks. It was also deployed over Tapi River crossing on 220 KV Kawas-Navsari Transmission line which became vulnerable to collapse due to flash floods, saving time of over six months and ensured availability of power from Kawas project. Recently, it was also used in Gujarat as disaster management caused by cyclone in June, 1998 where the transmission system of about 17,000 kms of high tension and about 14,000 kms of low tension apart from other distribution systems was severely damaged. POWERGRID within few days restored the transmission lines through ERS to facilitate evacuation of power from generating stations to important cities of Gujarat.

Project Management

A high level in-house engineering and project management team enables POWERGRID to execute its projects on time even in extremely difficult terrain and many a time in the face of insurgency and law and order problems. Recent examples in this field include the completion of the first 400 KV line in Jammu & Kashmir ahead of the commissioning of generating units and a host of activities including augmentation of systems in the North-East Region. Recently, it has commissioned 2×500 MW Chandarpur HVDC Back-to-Back System and Vindhyachal additional transmission system in Western Region on schedule as per the commitment made to the Government and a 400 KV Jeypore-Gazuwaka transmission line was commissioned 9 months ahead of schedule.

Contracts Management

A strong and effective procurement strategy is needed for providing effective support for implementation of such massive projects. The immense expertise garnered by

POWERGRID in this area can be gauged from the fact that Rs. 7,000 crores worth procurement has been made in time. Moreover, the model bidding document prepared by POWERGRID for ICB has now been followed by many utilities in the power sector. POWERGRID has made concerted efforts in vendor development and has supported BHEL for supply of insulators to be used for 800 KV AC/HVDC systems. International Institutions like the World Bank and the Asian Development Bank have time and again lauded POWERGRID's achievements in this area.

Complements from International Consultants to World Bank

"Engineering Specifications, Contract and Project management, Contract documents has been of excellent quality"

Construction Management

On the construction front, POWERGRID has been accomplishing all the committed targets during the past six years as per the commitment made to the Government for timely evacuation of power from generation stations—a fine example of the success of its integrated project management practices. It has added about 11,000 ckt kms. EHV transmission line along with 17 sub-stations with a capacity of 16,000 MVA. More than 20,000 ckt kms. of transmission lines are under construction during next decade. In recognition of its deep commitment to quality consciousness POWERGRID has been awarded the prestigious ISO 9001 Quality Systems Certificate, by NQA—Quality System Registrar UK, for turnkey execution upto 400 KV transmission system. POWERGRID thus gained the distinction of being the first Indian power utility for such accredition.

Operation and Maintenance

POWERGRID employs state-of-the-art thermo vision scanning, condition monitoring of EHV equipment, hot-line maintenance techniques, etc. to ensure maximum line availability without interruptions even during maintenance activities which reflects operational and maintenance expertise with transmission system availability being maintained over 98%. POWERGRID has also introduced a special earthling device in case of double circuit transmission line for maintenance of one circuit when the other is charged.

Complementing POWERGRID, the World Bank has stated that

> "The availability of the POWERGRID, system is over 98%. System expansion is progressing at a fast pace, and being implemented in a highly satisfactory manner"

Business Development and Consultancy

POWERGRID has made great strides in the consultancy area to provide its experience and expertise to other power utilities and today executing orders worth more than Rs. 30 crores. The service offered by POWERGRID include power system studies, engineering and turnkey execution of transmission systems. Few major clients of POWERGRID worth mentioning include Merz and McCellan, U.K., State Electricity Boards of Orissa, Bihar, Pondicherry, Andhra Pradesh Karnataka and West Bengal, Delhi Vidyut Board and West Bengal Power Development Board.

The Power Minister, West Bengal Government has lauded POWERGRID's endeavour and stated:

> "Here is an organisation that meets its commitments and values time. I wish there are more organisations with a similar outlook"

Human Resources

POWERGRID's human resource consisting of over 7,000 professionals is spread all over the nation is truly representative of the strength and philosophy of the country i.e. Unity in Diversity. POWERGRID believes that its crore strength lies in its people. POWERGRID's institutional development and growth has been impressive. It has amalgamated staff into a surprisingly homogeneous team.

DEVELOPMENT OF NATIONAL GRID

As mentioned above, the energy potential in the country is concentrated in few pocket, thus high priority is being given to integrate all the regional grids to form a national Grid, so that the energy requirements could be met on a national basis. This work is being carried out by POWERGRID under different phases. In the first phase, considering the wide variation in grid parameters of different regional grids, HVDC back to back inter-connections have been planned to interconnect all the regions. Two such interconnection between Northern and Western regions, and Western and Southern regions are already under operation. Another two links between Southern and Eastern regions, and Eastern and Northern regions are under different stages of execution and would be ready by 1999 and 2001 respectively.

POWERGRID's vision to integrate the regional grids which would be implemented by the turn of the century, would establish much needed basic infrastructure for a National Grid facilitating transfer of 1000 to 1500 MW of power from any one region to another. With the operation of National grid, we can also aim towards levelised peak, as the peak load of one region can be met by the neighbouring region as the peak load of each region also occurs at varying times, almost at a difference of an hour since the country lies between the latitudes of 73° (Calcutta) and 89° (Bombay).

The interconnection between regions would be further extended through high capacity links which have been planned to evacuate power from large sized projects in Eastern region viz. Talcher-II (2000 MW) and Ib Valley (3960 MW) in Orissa. These links envisage construction of high capacity HVDC bi-pole (2000 MW) between Eastern and Southern regions and another HVDC bi-pole (3000 MW) between Eastern and Northern Regions.

During 10th and 11th Plan periods, as a long term perspective, EHV AC links (765 KV or higher) have been envisaged to connect Eastern and Northern regional grids while interconnection to Southern region would be strengthened through additional HVDC bi-pole. Besides, high capacity EHV lines in Chicken Neck area have been planned. These long term schemes have been envisaged to evacuate power from the major hydro projects expected to come up in North-Eastern region, Sikkim and Bhutan and coal-based projects in Orissa and Bihar.

INTEGRATED SYSTEM OPERATION—UNIFIED SYSTEM CO-ORDINATION

POWERGRID is implementing unifield System Coordination and Control project (SC&C), encompassing control centres and associated communication facilities in each State and Region through state-of-the-art SCADA and EMS technologies. This would facilitate real time monitoring and date acquisition of the power system for optimal utilisation of the national resources to provide most economical power to the consumer with higher standards of reliability and security. This will help consistency in design, implementation, maintenance, compatible equipment and system and fully dedicated integrated communication network throughout the nation. The implementation of these schemes is in progress in Northern and Southern Regions and by the year 2002 the Unified SC&C facilities will be in operation all the regions. While the transmission links provide requisite hardware to facilitate flow of power, system co-ordination and control schemes would facilitate integrated system operation aiming at economical and commercial exchange of power with enhanced security and reliability.

Financial Resource

POWERGRID envisages an investment of about Rs. 13,000 crores for its identified on-going and new schemes during Ninth Five Year Plan and possibly enhance it by an additional investment to the tune of Rs. 5,000 crores for providing evacuation arrangements associated with mega IPPs. Against Rs. 13000 crores of investment more than Rs. 10,000 crores has already been tied up and balance funds would be met from internal resource generation and domestic borrowing. The funds so far tied up includes various multilateral and bilateral agencies like the World Bank, Asian Development Bank, OECF, J-Exim, European Investment Bank, Syndicated etc.

The strong investment programme planned by POWERGRID will necessitate large scale borrowing, for which capital restructuring by converting existing loans from Government of India of about Rs. 1,200 crores into equity is envisaged. This capital restructuring would not only improve its capital base but would also enable POWERGRID to borrow from multilateral agencies, international and domestic financial markets. Other financial strategies include its recognition and rating by international agencies to raise financial resources through GDRs, Syndicated loans, etc. and re-financing of its costlier loans which it has inherited from the transferred organisations through much cheaper loans in the prevailing market. The Financial strategies ensure that POWERGRID will not need any budgetary support from the Government of India and the company will be able to meet its committed investment programme to the nation.

RATIONALISATION OF COMMERCIAL ARRANGEMENTS

Present Tariff Structure

Presently the central sector PSUs are supplying power to the SEBs based on a flat rate tariff (paise/Kwh.) Even inter-State and inter-regional exchanges of power are based on this structure. This does not encourage merit-order operations of the power plants. Thus, even the pithead station owned by the Central PSUs are asked to back down inspite of having a lower short run marginal cost, while generation continues at load centre stations with higher short run marginal cost.

New Availability Based Tariff Structure

To remove the anomalies in the present tariff structure, Government of India in November 1996, decided to implement Availability Based Tariff (AT) for the generating stations as well as future IPPs along with decentralised dispatch by SEBs with RLDCs coordinating regional and inter-regional operations i.e. operating loose regional power pools as the operating mode for the regional grids.

The proposed availability based tariff has two parts structure based on generating unit availability. The first component is the 'Capacity charge' to cover the total fixed cost of the generating and the second component is 'Energy Charge' component which is to be paid by SEBs to the generators as per the SEBs scheduled drawal. This charge is meant to cover the variable cost which is primarily the fuel cost of generating.

Generating companies can meet dispatch orders only by regulating their output of generating stations and if they fail to meet the schedule, the deviations would have to be covered by unscheduled interchange. This Unscheduled Interchange (UI) is going to be the frequency based pricing in the proposed tariff system. The implementation of this new tariff structure has already commenced in Southern Region wherein a mock exercise has been successfully completed and will be fully implemented soon. In phased manner the implementation in other regions will be completed by the year 2000.

EXCHANGE OF POWER FROM NEIGHBOURING COUNTRIES

India as a major partner in the South-Asian region, can play a vital role in developing a sound power sector in the overall region. In this direction, India has been assisting Nepal and Bhutan for past many yeaRs. Presently, India and Nepal are exchanging power to a limited extent at different points. Bhutan is exporting bulk of the power from Chukha project, implemented with Indian technical and financial assistance. The prospects of jointly developing a number of new large sized projects in Nepal (e.g. Pancheswar—6,000 MW, and West Seti HEP—750MW) and in Bhutan (e.g. Tala—2,000 MW) are under consideration from which a major portion of power would be supplied to India.

It is reported that Bangladesh has vast resources of natural gas in Eastern sector which need to be developed for power generation which could be exported to India on one side any Myanmar and Thailand on the other.

PRIVATE INVESTMENT IN TRANSMISSION SECTOR

The transmission sector would require Rs. 57,000 crores during the Ninth Plan but adequate funds are not available with either POWERGRID or SEB's. Thus, it is imperative to invite private sector participation. However, transmission not being identified as a distinct activity as per the ES Act, private investors would not be able to enter this sector and as you are aware a Bill in the Parliament has been introduced to identify transmission as a distinct activity independent from generation and distribution to enable Private Sector participation in transmission.

CONCLUSION

POWERGRID will be instrumental in constructive restructuring of the Indian Power Sector and will continue to act as a change agent to motivate and catalyse development in all areas of power generation, transmission and distribution and will extend and share it benefits to the neighbouring SAARC nations for formation South Asian Grid for mutual benefit of each other. It has committee to introduce a culture of competition for infusing efficiency and improving the quality of service by well established Power Pools.

It will strive to for creating a level playing field for existing SEBs and generating companies and will provide easy access to new small and large IPPs so that market oriented mechanism can thrive on competition with self sustaining growth to sell their output to the viable SEBs, private distribution companies and bulk consumeRs. However, in view of the emerging scenario of complex and sophisticated bulk power market a strong independent regulatory body is needed which should enhance the development of market forces for reliable, efficient and economic service goals. The formation of POWERGRID is only a tip of the iceberg and bigger and greater changes are in the offing in the Indian power sector for building enduring institutions for a sustainable future.

PART-III
OVERVIEW PRESENTATIONS

FINANCING THE GLOBAL ENERGY SECTOR : THE TASK AHEAD

9

A.A. Churchill
President of the Washington Energy Group,
Washington (USA).

The Issues

At the 16th WEC Congress in Tokyo (1995), concerns were expressed on the adequacy and availability of the capital resources required to finance the growing needs of the global energy sector. The objectives of this Study are to address these concerns and to suggest ways to overcome the threat to economic growth from potential financial constraints in the energy sector.

The Message

> **The first conclusion of the Study is that global capital resources in principle are more than adequate to meet any potential demands coming from the energy sector.**

These demands are unlikely to exceed 3-4% of global output, the same proportion that has prevailed over past decades. Economic growth and the use of energy are closely linked, particularly in the developing world, and global economic growth will produce the necessary resources to fund energy investments. This is confirmed by the high savings rates found in most developing countries. Contrary to popular belief, savings rates in many developing countries are double those of the US and generally one-third greater than those of Europe or Japan. The experience of India is illustrative of the trend in saving in most major developing countries (see Figure 9.1).

International capital markets are growing rapidly and the call on these markets for energy investments will remain a small fraction of total market transactions. The major conclusion of the Study is that, while there is no overall or global constraint on capital resources there is a real problem in mobilising these resources for energy investment, especially in those countries which do not satisfy investors's criteria.

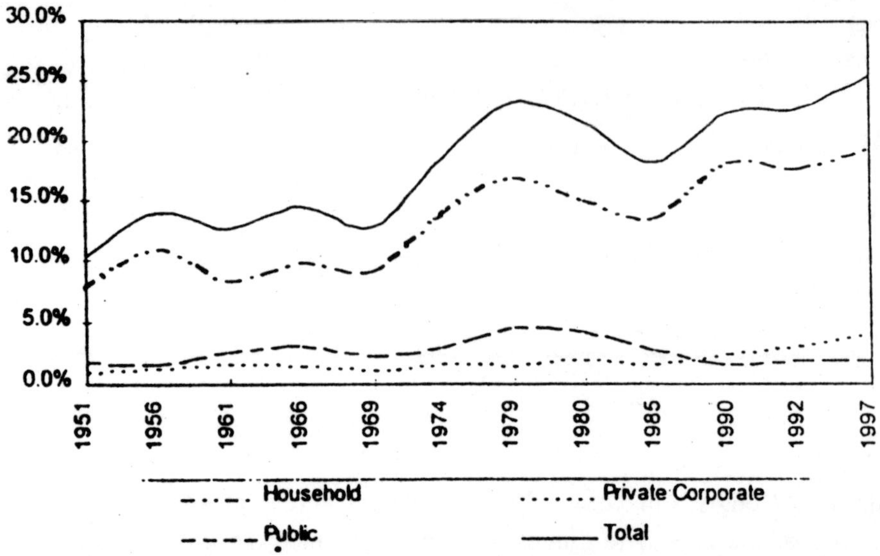

Fig. 9.1 : India : Gross Domestic Savings at Current Prices (% of GDP)

The Forces at Work

The Study starts with an overview of the links between economic growth and the use of energy. In the coming decades, over 90% of the growth in energy demand will come from the developing world. In many of these countries, rapidly rising incomes, the shift from rural to urban areas, and dramatic increase in industrialisation and mobility, will result in several decades of energy-intensive growth. In contrast to the developed world where energy demands and growth are less closely linked, such countries are likely to see energy requirements growing more rapidly than national incomes.

By the end of the period considered in the Study (2020), many of these countries will probably have made the transition from energy-intensive growth—driven by the demand for industrialisation, motorised transport, refrigeration, air-conditioning and all the accoutrements of modern life—to patterns of growth more closely resembling today's industrialised countries.

In examining the potential demand for energy in the decades ahead and the likely capital requirements for energy projects, the Study limited itself to reviewing existing scenarios, particularly the World Energy Council's (WEC) scenarios contained in *Energy for Tomorrow's World* (1993), and the recent Study Report, *Global Energy Perspectives to 2050 and Beyond* (1995) jointly undertaken by the World Energy Council and the International Institute for Applied Systems Analysis (IIASA). This Study subsequently had the advantage of access to more recent data and feedback from many of the Member Committees of the WEC.

On balance, the review, suggests that overall energy demand could be greater than the demands found in most of these scenarios. While this may raise concerns with respect to the global environment, it would primarily be the result of rapid economic growth in some of the world's poorest regions. Without this growth, much of the world's population would be condemned to unacceptable levels of poverty and deprivation —a situation which, unless dramatic change occurs, may nevertheless persist in some regions.

GLOBAL CAPITAL NEED AND RESOURCES

In the WEC Commission Report *"Energy for Tomorrow's World", the* Reference Case B assumed for the period 1990-2020 a cumulative energy supply investment requirement in 1990 dollars of US$30 trillion ($10^{12}$) or roughly 3-4% of global GDP. This is in line with historical experience and provides a sufficient degree of accuracy for the purposes of this Study.

There are other scenarios, notably those contained in the Study Report *Global Energy Perspectives* to 2050 *and Beyond,* which suggest a decline in the capital needs of the global energy sector relative to income. These scenarios are based on a more sophisticated analysis of changes in energy intensities resulting from changes in both technology and the structure of the global economy. Once adjustments are made, however, to take account of possible efficiency improvements related to environmental expenditures and more effective use of capital, the scenarios of these two studies are consistent. This Study proceeds on the basis that it is more realistic, and perhaps safer, to assume that investment requirements will remain at approximately the same levels of GDP as they have been in the past. Implicitly, the assumption is the that the combination of increased demand for energy intensive goods and the rapidity of structural changes in many parts of the developing world are likely to offset improvements in energy efficiency. This is particularly true if, as it presently appears, the world of the next two decades will continue to move along a substantial growth path with low energy prices and relatively few policy or supply constraints. By 2020, the annual investment requirements could be as high as US$1207 to US$2118 billion or 3 to 4% of world GDP.

The major issue is how to mobilise capital. The capital mobilisation problems faced by many countries are not affected by assuming lower or higher levels of growth. What is affected is the ability to harness local capital and the sense of urgency with which the existing constraints need to be addressed. Unless action is taken quickly, potential bottlenecks in the energy sector could become a limiting factor to economic growth and to the movement out of poverty for almost 40% of the World's population.

Most of the pressing issues in many developing countries are found in domestic markets for energy services. International capital flows, both private and official, can support domestic efforts to raise resources, but they are seldom substitutes.

The energy sector and, in particular, the requirements of electric power are too large relative to the overall size of national economies to be dependent exclusively on external savings. The need to develop domestic capital markets is as urgent as the need to develop the energy sector.

THE REGIONS

There are unique challenges associated with each region. To link to global capital markets, each region must balance domestic concerns within its own institutional framework.

The Least Developed Countries

Most of the difficulties of energy financing have special force in relation to the least developed countries. One of the Study's greatest concerns is that two billion people in the developing world have yet to be reached by commercial energy and have little prospect of this occurring unless fundamental measures are implemented. In some of the least developed countries, the political, legal and institutional structures necessary to underpin conventional investment are non — existent. These countries are not susceptible to the financing solutions which work in countries which are structured differently: they often have no capital markets to be made more efficient — their utilities are often desperately short of capital — their customers often cannot afford to pay — their authorities often lack the capacity to borrow, and do not appear to have the will to implement change.

The least developed countries can also be the most susceptible to expensive and inappropriate solutions. They may lack the ability to discriminate amongst the technical solutions on offer. They may not know how best to use the limited funds that they may have. In addition, some of these countries are easy prey to corrupt proposals.

Although the least developed countries are the object of special consideration by the multilateral development banks and similar institutions, they also require private sector assistance to design structures and mechanisms for reducing investment risk and to arrange funding.

Rising energy demands, declining public investments and the evolving role of the multilateral financial institutions—from that of lenders to public enterprises to that of facilitators to private investment—require increased efforts by governments to change their roles to enable them to create a business environment attractive to private investment both international and domestic.

Africa

Although Africa is impressively endowed with energy resources, the development of commercial energy is constrained by the generally poor state of economic development, inadequate infrastructure, external indebtedness, and by political and civil unrest. Africa with 13% of the world's population currently only consumers 3% of its commercial energy. The problems of Africa go well beyond the energy sector and its financing and there are no easy immediate solutions.

Rapid population growth already places considerable additional pressure on African societies. One ground for optimism is that addressing Africa's energy needs requires a relatively small share of global resources. The real challenges are the ability of African nations to undertake dramatic and radical reforms and the willingness of the rest of the world to provide the necessary support.

Closer co-operation and integration within the Southern African grid, which have had significant positive effects on the overall need for resources and their financing, provide a good illustration of how problems can be addressed on a regional level. This is a model that needs to be considered in other parts of Africa. Most African countries are small in terms of their energy sectors and few private investors are willing to put resources into overcoming the difficult institutional environment in each country. Greater trade and co-operation in primary energy and electricity can ease the financing burdens.

Energy investments in Africa should have a higher priority in the World Bank and other multilateral and bilateral agencies, since apparently in other regions there is greater ability to undertake the type of reforms needed to attract both domestic and foreign investment. The agencies should direct their support towards restructuring and unbundling the inefficient state monopolies and to setting the framework for more effective private participation.

The Middle East and North Africa

This is a diverse and rapidly growing region. Energy intensity is static and the needs for desalinisation and air-conditioning will continue to drive demand. The main financing issues are the need to develop domestic capital markets and to rely less on state funding for energy investments. Significant investment will be needed to develop oil production and refining capacity.

Latin America

Latin America is a region where the process of reform is well under way and where the ability to undertake reforms has been the greatest among developing regions. In most

Latin American countries, governments have gradually reduced their direct participation in the energy sector and have focused on the formulation of policies and regulations. Although there is still much to be done and not all countries of the region are at the same level, there is a growing confidence within the region that it can mobilise the resources it needs to meet its energy requirements.

South and East Asia

It is in the countries of Asia that much of the present and future growth in energy demand is concentrated. There is also great diversity among the countries.

The major financial impediment in this region is the inefficient and extensive involvement of government in the energy sector. Besides direct state ownership of energy facilities, state interventions include credit allocation via national banks, distortional interest rate and pricing policies and interest rate and foreign exchange controls. Government dominance has stretched demands on public funds beyond sustainable levels and this has impeded the development of both private capital markets and the energy sector. The energy requirements of South and East Asia are too large to depend on external financing for more than a fraction of their needs.

Domestic savings in this region are over one trillion US dollars a year and are clearly adequate to meet the annual US $ 60 to US $ 100 billion in energy financing needs over the next decade. The challenge for most of the countries of this region is institutional in nature; can they undertake difficult political and institutional reforms in the energy sector in time to prevent the availability of energy becoming a constraint on development?

In order to address the demands generated by rapid growth, most of the countries in Asia are indicating an increasing willingness to utilise private capital. Although state enterprises still maintain a dominant role, this is being tempered as experience is gained with private sector alternatives.

The Study has singled out two countries, India and China, for closer examination. Together these countries will possibly represent almost half of the future increase in world energy demand. Both have formidable political, ideological, and bureaucratic constraints in their energy sectors and overcoming these will be a major challenge on the path towards further development and improvement of efficiency in these sectors. Mobilisation of mainly local capital for energy investments will be the key success factor for the economic aspirations of these two countries. Although China's and India's energy sectors may appear to poste a global financing problem, solutions to their problems can realistically be mainly domestic.

The international community can and should provide help in the search for solutions, but it cannot dictate them. The countries themselves are beginning to realise that, unless there are significant changes in the role of the state, financing energy investments will remain a problem. The unleashing of the obvious entrepreneurial skills prevalent in their populations will most certainly make a dramatic difference.

Central and Eastern Europe and the CIS

The economic-in-transition in Central and Eastern Europe (CEE) and the Commonwealth of Independent States (CIS) are all substantial energy users. Some have most of the natural resources required to meet their future energy needs. Overall investment requirements between 1990-2020 estimates at US $ 3 trillion are, however, constrained by the pessimistic assumptions of economic growth. Some caution is required in projecting present trends too far into the future; these are economies where household consumption has been artificially constrained and the unleashing of market forces could see a dramatic rise in consumer demand for energy intensive goods and services.

The need for rapid political and institutional reforms is critical in CEE and the CIS.

Presently, flows of international capital for energy investments into this region are modest and concentrated in countries with accelerated reform processes or exportable energy.

Domestic capital markets are too small and too short-term oriented to cope with the needs of the energy industries.

Like the large countries in Asia, the CEE and the CIS are in the process of redefining the role of the state and beginning the transition to more market-based economies. Apart from the export side of the energy business, they are unlikely to receive significant inflows of capital without considerable and rapid reforms. The issues are political and institutional rather than financial.

Western Europe

Western Europe is currently moving from a predominance of state institutions in the energy and financial sectors towards an integrated European market. This transition is accelerated by the competitive pressures introduced by globalisation, and the increasing demands on public funds. The United Kingdom and the Nordic countries are well advanced in moving away from monopoly and introducing competitive markets. Other countries are struggling with the potential social costs of reform. Excess capacity in electric power is likely to be an additional problem. Despite extensive government participation in the energy sector, financing is not an issue in Western Europe because financial and legal institutions are well developed.

North America

The North American energy and capital markets are the largest and most developed in the world. Almost all financing of energy is provided by private capital. Even where public institutions are responsible for energy investments, they have used private markets to raise most of the funds. The major change taking place is the shift from regulated private or public monopolies to competitive markets. This is introducing dramatic changes in the industrial structure of all energy industries.

Unbundling, mergers, convergence and other changes are redefining traditional concepts in ways that are still being unveiled. Many of the pressures for change are in response to over-capacity and technological progress as well as to global trends. The precedents for restructuring the industries of many countries will emerge from this experience. US firms, with their access to large capital markets and their experience with competition in domestic energy markets, are also in the forefront of global competition.

THE FUELS

The energy production and consumption chain is complex in its financing; investments are required to develop primary fuel sources, to transport them to market, and to convert them into useful energy services. Each fuel source has unique aspects when it comes to financing.

Oil

Traditionally, oil has been a profitable investment with large economic rents available for sharing between international companies and host governments. Retained earnings have accounted for the major share of new investments. Over time, as capital markets have grown and become more sophisticated, large oil companies have increasingly used their substantial balance sheets to raise investment funds in these markets. As a result, investments in hydrocarbons have become more closely linked to the overall return on capital.

From a global perspective, there appear to be relatively few financing problems in the upstream oil sector. From 1991 to 1995, investments in exploration, development, and production averaged about US $ 80 billion per year world-wide. This is expected to increase to about US $ 100 billion a year over the next five years. The fact that oil is an internationally traded commodity with sophisticated commodity market support mechanisms eases the risk sharing and financing process.

Major financial issues for the sector are:

- The balance between returns on capital and host countries interests a delicate matter reflecting judgements on risks and rewards. There is always the danger, in the bargaining process, of the various parties making mistakes of judgements.

  A number of countries for political reasons, have limited the access of international companies to their markets. They have chosen national state companies relying on public funds which may not be adequate to meet future investment needs.

  China and India are of particular concern in this respect. In order to overcome the political risks in some countries, there have been suggestions that the World Bank or other public bodies should assist by taking on the political risk. This would be an ineffective use of these sources of finance. Oil is a mature industry, well able to evaluate and manage risks on its own. In most cases, lowering perceived country risk is in the hands of the host governments.

  To those countries reluctant to open up to international capital, the Study suggests considering emulating Indonesia, Saudi Arabia and other oil-rich countries which have developed sophisticated contractual arrangements to allow them to benefit successfully from international participation in the industry without either compromising rents or national sovereignty.

  National oil monopolies can also present financing problems, since State ownership means the government is responsible for financing. Where these monopolies are primarily domestic enterprises importing, refining, and distributing the product, they are often of questionable efficiency and, as in the electric power sector, there are strong arguments for introducing private ownership and competition. Where they are export-oriented as well (e.g. Venezuela, Mexico), their ability to generate investment funds can be compromised by the government's need for revenues.

- Pricing policies which are market-oriented help considerably in solving most financing problems.

  In some least developed countries, however, market solutions cannot overcome, chronic problems such as political risk. Some type of industry-aided funding initiative should be considered.

Gas

The demand for gas is expected to grow dramatically in most countries. While world supplies are more than adequate, distribution requires capital intensive delivery systems either in the form of pipelines or LNG facilities. The WEC estimates range between US $ 900 and US $ 2600 billion in cumulative investments over the period 1990 to 2020. A critical size is necessary (usually from the power sector) for investment in gas delivery to make economic sense. In many countries, infrastructure, markets and government pricing policies are simply inadequate for profitable gas delivery. Participation of more international investors will provide key skills and enhance financing in this sector. Where governments are responsible for both investment in electric power and gas infrastructure (e.g. India and China), this can result in difficult, if not impossible, problems or raising adequate public finance. In countries, where electric power is a commercial business (e.g. Argentina, Chile, Bolivia), the private sector appears willing to undertake the necessary long-term contractual arrangements to develop and finance the infrastructure. Where gas enters international trade either through pipelines or in the form of LNG, large amount of capital have been raised in international markets to support gas projects.

Gas demand has been growing rapidly in the past decade and therefore future projections of demand may be understated. Demand growth in Asia, for example, is constrained by present perceptions of limited supplies within the region. Investment in exploration and improvements in trade policies could result in significantly greater demand, particularly for electric power plants and industrial use.

The exploration, development and transport of gas usually requires significant up-front investment. Close co-ordination between investments in the gas and power infrrastructure in required. In the well-develolped markets of Europe and North America, contractural arrangements between suppliers and users meet this need, but in less developed markets there are a number of obstacles to these market-based arrangements :

- The gas industry needs to be opened up to a broader range of investors. Where national monopolies are responsible, for both oil and gas (e.g. Mexico, Brazil), they have tended to focus on oil as the more profitable side of the business at the expense of gas. Allowing the power comprnies and other industries to invest in gas can increase the availability of supply.

- Where gas is in the hands of state gas monopolies (e.g. India, China,) eliminating monopoly and introducing competition needs to be considered. As long as the State is responsible for financing and public finances are inadequate, there is the danger that supply will be constrained.

Coal

Ownership of world coal production is roughly half private and half public. Coal is mainly a domestic commodity, with only 15% of the world's output entering international trade. There countries account for most of the world's trade in coal : South Africa, Australia, and the USA.

Similarly to oil, where coal is traded internationally, financing, is not a problem. In the traditionally, financing is not a problem. In the traditional coal exporting countries, coal is mined by large companies capable or raising the necessary capital. Newcomers into the international coal market (e.g. Indonesia and Colombia) also have been to raise the necessary funds.

The financing difficulties for future coal production are largely confined to the CEE/CPS, India and China, where large public monopolies in production and transport dominate. Output is sold at prices that do not always covers cost, efficiency is far from international norms and environmental degradation is a serious issue. Between 1990 and 2020, world coal consumption is estimated to increase by 716 million toe, requiring approximately US $ 4 trillion (or US $ 13 billion per annum) investment. Given the existing and potential rates of economic growth in India and China alone, this number is probably seriously understated.

Financing problems extend beyond the mining of coal and include its preparation and transport. They also have a negative impact on electric power. For every dollar spent on expanding coal capacity, approximately US $ 15 is spent on expanding electric power. The poor quality of coal and its delivery systems can threaten the productivity of the much larger investments in electric power.

Given the importance of coal as a primary fuel, a high priority will have to be given to financing new capacity and improving its efficiency and environmental performance. The options are limited :

- The price of electricity will have to be increased to the point where it is able to pay prices for coal that cover the costs of improving and adding to existing capacity. Both the railroads and the mines will have to generate sufficient earnings either to finance these requirements or to have sufficiently high returns to be able to attract private capital. Participation of more international investors will provide key skills and enhance financing in this sector.
- A more efficient institutional structure is required for coal mining. The present state enterprises are inefficient and have difficult labour and social problems. The

unbundling of the monopolies and a broader participation of the power and industrial sectors will be necessary.

- Modern mining methods need to be introduced. Some level of international investment will be necessary to accomplish this goal. This will require substantial changes in the existing incentive structure, particularly the relationships between the mines, the railroads and the power companies.
- Privatisation of the coal mining sector needs to be considered.
- Transport policy needs some rethinking. In India, for example, the lack of investment in coal preparation adds significantly to the tonnage of materials moved. Although both India and China are improving the location of power plants with respect to the mines, there is still considerable room for improvement. A more efficient transport and location policy will lower investment requirements.

ELECTRIC POWER

The biggest financing challenge both in relative and absolute terms is in electric power. It is larger than all other energy investments combined. Public monopolies dominate, prices are politically determined, and substantial. In the past it has relied on public revenues to support most investments, and regulatory and financial mechanisms are poorly developed. Under these conditions, financing can be a problem. This industry is in the process of dramatic change, partly in response to financing problems but also in response to global pressures for improvement inefficiency. The on-going restructuring of the industry has implications for both developed and developing countries. Even in North America, where the sector has been dominated by privately owned and financed utilities, the opening up of the regional monopolies and the introduction of competition is profoundly altering the way the business is financed. Western Europe is still debating changes in the industry but, in the end, it too will follow global technological and market changes.

One already obvious outcome of these changes is the rise of international power companies. These companies, mainly from the USA and the United Kingdom, but to certain extent also from France, Germany and other countries, are undertaking the role of developer and financier of electric power in countries all over the world.As more countries restructure their power businesses, their leading domestic firms are also likely to enter the international market place.

Most countries, however, are a long way from arriving at the point where they can depend on private resources to finance the electric power sector. Some have ideological hurdles to overcome; others are still unconvinced of the need to change. Most have taken only tentative steps to restructure their industry in ways to permit greater private participation. In almost all cases, the need for restructuring goes well beyond the electric power industry. Given the pervasive role of the state in all aspects of economic activity, it is difficult, if not impossible, to make changes in one sector alone. Reforms across a broad front are required.

If private finance is to flow at a greater rate to the electric power sector, ways must be found to encourage private sources of finance to assume higher risks. There is obviously a limit on how far the credit standing of a government can be used to support private capital flows. If higher risks are to be taken by the private sector, greater control over its investment must be given to the sector as well. Many countries are trying to introduce private finance and ownership by encouraging independent power producers as an alternative source of new generating capacity. Instead of the state enterprise contracting and building new capacity, independent producers build and finance new plants on the basis of a long-term power purchase agreement. In some cases these contracts are subject to a bidding process and in others they are the result of direct negotiations between the utility and the IPP. The utility takes all of the market risk (that there will be buyers for the power at prices that cover costs) and the IPP takes the financing, construction, completion and performance risks.

In spite of the flurry of activity—latest reports indicate over 1000 Memoranda of Understanding between developers and governments—the results to date are disappointing. This type of financing has proved to be slow and expensive and only a limited number of projects of those under consideration ever reach financial closure. Projects are languishing especially where market-based prices are not permitted.

Although the financing benefits provided by the IPP approach are modest, and likely to remain so, IPPs have served as a useful first step in introducing private capital into the power sector. However useful IPP project financing can be in a particular case, it is not the only and by far not the most important financing mechanism for utilities. In Western Europe, for example, utilities whether public, mixed or private have no difficulty in obtaining loans or issuing bonds on the domestic or international capital markets.

Whatever the specific circumstances in the various countries, financing problems could be reduced by measures such as:

- **Enhancing domestic capital markets**
- Energy investments are a major part of the asset-base in capital markets at the early stages of development characterised by the lack of quality investments.

Provided the energy enterprises earn reasonable returns and are prudently managed they can become an important asset in these markets, particularly for institutional savings. Malaysia and Chile are examples of countries that have effectively used their capital markets to mobilise funds for the energy sector, while the economies-in-transition have so far been unable to mobilise long-term funds.

- **Accessing international capital markets**
- Utilities should familiarise themselves with international capital markets.
- **Enhancing public-private partnership**
- In circumstances where a purely commercial solution cannot (yet) be implemented, public-private partnerships can reduce financing problems or pave the way for subsequent solutions.
- **Restructuring tariffs, reducing subsidies**
- As long as the electric power industry operates at a loss or requires substantial state subsidies, it will be unable to attract private funds without government guarantees.
- **Restructuring the power industry**
- While the above measures often financing problems, the liberalisation and restructuring of the electric power industry is likely to offer a lasting solution under circumstances which utilities are unable and governments unwilling to inject or subscribe to finance.

In addition to the general financing problems discussed above, high front-end capital needs and other issues affect financing of hydro and nuclear power.

Hydropower

The main problem with hydropower lies in the nature of the public procurement process. In the construction of dams, politics and employment have played a greater role than cost controls and fiscal accountability. Even turnkey contracts turn out to be cost-plus and a large percentage of projects end up in arbitration or the courts before final payments are made. Private developers and their financiers regard large hydropower as a excessively risks investment, while small, run-of-the-river developments can more easily attract private finance.

In order to overcome both the constraints and excessive costs of public procurement and the private sector's perceptions of risk, new methods of building, managing and operating hydropower facilities will be needed.

- All parties participating in these projects need to develop mutual risk and profit sharing arrangements similar to those found in independent power projects. In the past almost all risk has been borne by the government.
- The risks (siting, construction, environment, resettlement) suggest the need for greater equity-to-debt contributions in the financing.
- There is need for greater entrepreneurship in developing hydro projects.

NUCLEAR POWER

High capital costs and, in some countries, a poor record of cost containment, together with a growing negative public perception, all result in increasingly unacceptable risks for private financing of nuclear power. Added to this are licensing and site location difficulties, waste disposal concerns and public vs. private liability issues.

Unless the nuclear industry itself is prepared to raise substantial equity capital to finance new plants and new technologies, nuclear plants will not be built without predominantly governmental financial support.

It nuclear power is to remain or become competitive, some of the following actions will need to be taken:

- Costs and construction times will have to be reduced substantially. Given the perceptions of risk, it will take fairly dramatic declines in costs to make nuclear attractive to private industry.
- Plants of smaller size and of modular construction would better fit today's market needs.
- The nuclear industry or government will have to finance and/or guarantee any initial new designs. In the face of growing competitive markets, the power industry is unlikely to take on this risk.
- The privatisation of waste disposal under enabling regulation may assist in overcoming the political problems associated with direct government responsibility.

THE POLICY ISSUES -REDEFINING ROLES

State Ownership

There are few issues as sensitive as the relationship between the state and the energy enterprises. State ownership can be, and often is, a major issue in mobilising resources for the energy sector. This Study has examined the question of ownership solely from the perspective of how it impacts on the ability of a country to raise resources for energy investment. State enterprises have been able to raise resources either through use of the state's power of taxation and state guarantees, or through their ability to earn sufficient returns to attract private capital. In the developed world, state monopolies have raised resources from capital markets because of the protections of the property rights of holders of financial instruments assured by a regulatory framework that produces the revenues required to service the debt. It does not, however, necessarily ensure efficiency—the major issue for developed countries is efficiency rather than capital mobilisation.

In the developing world, most of the capital resources are raised through the government's powers of taxation either by transfers from the budget or by utilising the state's access to domestic and international credit. Regulatory structures are weak or non-existent, and political rather than financial criteria control decision-making. In addition, public utilities function as the social arm of the government—to expand power supplies with governments as owners and operators. Under these circumstances, the enterprises are dependent on the state for resources; private savers are unlikely to put their capital into loss making enterprises subject to often arbitrary and unpredictable political decisions.

> **As long as the state is willing to underwrite energy enterprises and allocate public funds to meet investment needs, state ownership is not of itself a problem. It is only when the state is unable or unwilling to provide for these needs that financing can become, and often is, an issue.**

In most countries, the many demands place on the state to meet a widening list of educational, health and other social and economic needs are straining its capacity to continue funding energy enterprises. The alternative of state enterprises operating in a regulated environment that attracts private resources from capital markets does not appear realistic in most developing countries. Capital markets are not sufficiently developed and changing the way state enterprises are managed requires an on-going effort that meets with little success, even in the developed countries.

If private capital is to provide the basis for funding in the coming decades, the issue of ownership will have to be addressed. This is particularly the case in Asia—most of the countries of Latin America are already moving away from state ownership. As long as the state owns energy enterprises, it is unlikely they will be able to attract sufficient private capital to fund their investment needs. In the transition, the state will have to continue to provide some support to these enterprises and prepare itself for a more demanding regulatory role as guardian of the public interest.

State Monopolies and Private Financing

A number of countries, particularly in Asia, have tried to introduce private financing of power projects while maintaining the central position of state monopolies. Private capital is invited to participate in building new capacity through Building Operate-Transfer (BOT) or Build-Operate-Own (BOO) schemes. These have had limited success. The state monopoly remains the only purchaser of power and since the owners of the new capacity lack direct access to customers, they will insist on the government underwriting the performance of the monopoly.

Guarantees are another way of utilising the limited credit of the public sector but governments are reluctant to provide these guarantees (e.g. China, India, Indonesia). Until this dilemma is resolved, private capital flows will continue to be linked to the availability of public finance.

Following the experience of the USA with independent power projects (IPPs), many countries have introduced incentives to encourage developers to undertake investments in capacity. The debt financing for these projects is raised on the security of project revenues without recourse to the assets of the sponsors. The results to date are disappointing. It is slow and expensive and only a limited number of those projects under consideration ever reach financial closure.

Projects are especially languishing where market-based prices are not permitted. The inadequacy of institutional structures and how to accelerate the flow of private capital into IPPs are further discussed.

One main problem is the lack of credit worthiness of the agencies contracting to purchase the power. Providers of debt financing are unwilling to lend without assurance from the government. In addition, the risks, including foreign exchange risks, are often too high to justify the typical 70/30 debt/equity ratios. More equity is required which may raise costs beyond those acceptable to many governments. Risk and risk management are not always well understood by governments and these misunderstandings continue to be a source of conflict with developers. Governments need to take on a more active role in lowering risks. The international developers are responding to these challenges by moving from the limited recourse financing model to financing structures which make use of corporate or balance sheet financing. They are also expanding their interests beyond provision of new capacity to the purchase of assets in all aspects of the business. The international power business is rapidly concentrating in the hands of a number of the more capable firms.

Role of International Financial Markets

Few countries are in a position to rely on international financial markets to provide the bulk of the financing for their energy sectors. Most energy related goods and services are domestically consumed and therefore they have to be financed from local sources. International developers can play an important role in providing access to international capital markets, introducing new technologies, management and financing techniques, particularly during the liberalisation or reform process. Their role in supporting and developing domestic capital markets can be of critical importance.

Unfortunately, few governments recognise the transitional nature of the initial foreign investment by effectively using opportunities to support the development of domestic markets.

Pricing

The way energy is priced can be a major constraint on the financial viability of most enterprises. This is particularly so with electric power. As long as electric power is treated as a social or political good, it will be impossible to attract private capital. The issue is not just the level of prices but the way prices are structured and changed.

Where prices are determined by regulation, the stability and predictability of the regulatory regime is essential. In many countries, however, pries are determined by a political process that is difficult to predict. In these circumstances, private investors are unwilling to enter the market without some form of government guarantee. Even in those countries where new, more predictabile regulatory frameworks are introduced investors will be cautious until implementation experience is accumulated. To overcome these problems, a number of countries have turned to market-based prices.

Addressing the Equity Issues

Many governments assume that addressing the efficiency issues will have an adverse impact on equity and economic growth. After all, most of the subsidies and price distortions were justified either as promoting greater equity or more rapid growth. Reform may bring rising prices.

There is little evidence to support these propositions. "Low" prices are often an illusion. In the cases examined in this report, the inefficiencies induced by price distortions more than offset any gains attributable to subsidies. Industrial consumers, for example, usually have to compensate for the poor quality of low priced supplies by investing in stand-by capacity; the poorest consumers are forced to pay high prices for alternative supplies because they are effectively rationed out of the "low priced" market.

In general, the recommendations of this report will lead to lower pries. There is a lot of room in almost all countries for improvements in efficiency which, in turn, will lead to lower real prices. In the short term, however, in some countries, the reform process will inevitably have an adverse impact on entrenched groups that have benefited from the present pricing policies. Balancing the long-term gains for the economy against the short-term losses of a few remains a major political challenge requiring strong leadership.

Subsidies

Although a few countries continue to subsidise some petroleum products (e.g. India, Venezuela etc.) and coal (e.g. Germany, Spain, the United Kingdom), the major element of subsidy is found in electricity prices. The estimated subsidy to electricity users in developing countries is over US $ 100 billion per year or equal to their annual investment requirement. Most of these subsidies encourage the use of grid-based power supplies and discourage the use of renewables and other forms of non-grid generation.

Governments subsidise for a variety of reasons, usually either to promote economic development or to support income redistribution. Experience suggests that subsidising electric power is not a useful means of achieving either objective. Governments always run out of resources long before they run out of recipients for subsidies. Administrative rationing of capacity is then required and it is inevitable that the poor will be rationed out of the system. They have no access to electricity and the subsidies invariably go to the urban middle classes. Carefully targeted, self financing schemes, such as that of Eskom in South Africa, appear to be more effective in dealing with income distribution concerns.

The Study examined two cases involving substantial subsidisation—Maxico and India. In both countries it proved difficult to contain subsidies once they were put in place and they rapidly grew to the point where they threatened fiscal stability. The experience of India and Mexico also illustrates the political difficulties encountered in trying to eliminate or reduce subsidies. Yet, until this problem is addressed private capital flows will continue to require government support.

Technology Transfer and Innovation

Financing the transfer of technology is often viewed as a substitute for capital transfers. A review of the experience suggests this is of limited usefulness. In this age of the Internet and global telecommunications, most technologies are readily available, particularly those of greatest use to developing countries. The global energy industries compete vigorously on the basis of their technologies and are prepared to supply them on competitive terms.

The real issue here is that technology is not a disembodied good but necessarily comes packaged with management and technical services. Almost all studies show that the productivity gains associated with technology come from the management and institutional changes forced by the technology rather than by its simple introduction. Private investment continues to be an efficient means of technology transfer and innovation.

Another major factor in effective technology transfer is the absorptive capacity of countries. This correlates highly with literacy and general levels of education. Countries like the Republic of Korea and Singapore have established educational and training systems that allow profitably to tap the international markets in technology.

Environment

Although the need for minimising environmental impacts can add to costs, it is not an overwhelming impediment. Various studies suggest that most environmental concerns can be accommodated within a 15-20% range of additional cost. For new capacity it can be as low as 10%. Provided environmental concerns are carefully prioritised, they should be present major financing problems. Benefits are both clear and conspicuous.

In many cases the appropriate environmental controls are built into new technologies. Most firms involved in investing in new plant and equipment are careful not only to comply with existing environmental standards but go further to protect themselves against costlier retrofits that will be necessary with the inevitable tightening of standards. The World Bank and other agencies have served as useful role in encouraging responsible environmental investment behaviour.

THE SEARCH FOR SOLUTIONS

In raising finance for the energy sector, there are major differences between countries and regions with regard to the characteristics and extent of their problems. There are also significant differences among the various segments of the energy sector. Trade, in particular, makes an important difference; if a fuel is a tradable commodity, it increases the possibility of tapping international markets to finance its producing. Solutions must necessarily be tailored to account for these differences.

As the Study notes, major issues how to mobilise capital rather than the adequacy of either the supply or demand for investment funds. In almost all cases, the nature of the problem is reflected in institutional structures and the potential solutions are dependent on adjustments in these structures. These structures invariably have a long history and reflect and political, social, and economic fabric of nations. They evolved or were put in place in response to perceived political, social and economic needs. Institutions are necessarily conservative in nature and have a degree of inertia in the face of pressures to change.

The world, however, is in a period of unprecedented change; the globalisation of production, new technologies, and the rising expectations of the developing world are part of this change.

The type of solutions discussed below are mainly about institutional adjustments required to meet the challenges of emerging national and international markets and their underlying needs for resources. Countries and their institutions will have to adjust to a process that if often painful and politically difficult.

There are no easy paths.

WHAT LESSONS CAN BE LEARNED FROM THE EXPERIENCE DETAILED IN THIS STUDY?

The lessons below must be seen in the context of a broader movement towards greater private sector participation and the shift from monopoly to competition. Not all lessons apply to all countries and all sectors. Their application in specific circumstances will vary depending on characteristics usually unique to either the country or segment of the industry. Most apply to electric power which presents the major financing challenge.

Improve the Efficiency of Domestic Capital Markets

While not specific to the energy sector, this lesson is fundamental to solving the financing problem. Generally, improvements in the legal and accounting structure along with prudent regulation are required in most countries. Capital markets will not attract domestic savings unless the savers are assured reasonable protection of their protection of their property rights. In addition, the development of mortgage markets, insurance and pension obligations can all provide an institutional source of savings.

Use Energy Investments to Assist in Developing Domestic Capital Markets

Energy and other infrastructure investments are a major part of the asset base in capital markets at the early stages of development. The problem with many domestic markets is the lack of quality investments. Provided the energy enterprises earn reasonable returns and are prudently managed, they can become an important assets in these markets, particularly for institutional savings. Malaysia and Chile are examples of countries that have effectively used their energy sectors to mobilise domestic funds.

Specific measures can support the link between capital market development and financing energy investments:

- Even when the state insists on majority control, it is useful to sell a minority ownership in capital market. It provides the market with needed assets, serves as an independent indicator of value, and acts as a brake on excessive political interference in management. This has been done, for example, in the Republic of Korea.

- Use the guarantee power of the state to promote liquidity in the capital market. The major problem in these markets is not the lack of long-term funds but rather the lack of liquidity of most financial assets. Using public resources to develop financial institutions that 'make' markets is an efficient use of limited public funds. These institutions, for example, can guarantee the roll-over or refinancing risk of shorter term bank debt—a major issue in financing most power projects.

- Rather than require foreign investors to turn over assets to the state or its agents (the BOT-concept), encourage investors to gradually divest ownership through sale of shares in the domestic capital market. This helps establish the market for financing future investments.

- Encourage foreign investors to seek domestic financial partners. The reputation of foreign firms and their visibility can provide comfort to domestic investors and enterprises as viable alternatives for their savings.

Transform the Power Sector into a Commercial Business

The transfer of ownership from the state to the private sector is a complex and demanding task. In addition to the financial issues, there are issues of equity and efficiency that need to be addressed. Fortunately, in recent years, a great deal of experience has been gained. A large number of countries in both the developed and developing worlds are making the transition. The developed world is largely motivated by the desire to improve efficiency and the developing world largely by the need to overcome financial constraints.

The process of privatisation offers a number of opportunities for addressing financial management issues. For example, asset values can be adjusted to reflect reality. In the

USA, where restructuring of the privately owned electric power industry is taking place, there are difficult problems in dealing with the issue of who bears the costs of excess capacity, now called 'stranded' or unnecessary assets. In countries that have moved from public to private ownership, for example, Argentina, Brazil, the United Kingdom and New Zealand, government 'equity' was adjusted to reflect the values of the restructured industry. Adjustment of asset values or the sale price of the enterprise has also been used to buy time to adjust prices.

Most public enterprises are not efficient in their management and employment practices, mainly due to government interference, some of the revenues from privatisation can be used to correct these problems. The restructuring in Brazil and Argentina offers good examples or how this can be done. There is also the issue of equity and fairness and how to address the society's need to provide access to energy to poorer segments of population.

Some of the issues of equity and fairness can be addressed through the privatisation process. In almost all cases in Latin America, the state has retained a significant minority ownership in order to share more widely the gains from privatisation. In Bolivia, the state's minority ownership was invested in the pension funds.

> **Where countries have privatised their industries and undertaken the structural adjustments necessary to privatise, the capital resource constraints have ceased to be significant.**

Use the Opportunities Provided by Privatisation to Increase Domestic Ownership of Energy Enterprises

In Bolivia, the government retained a minority share in the power companies which was then turned over to the pension funds as a substitute for direct government contributions. In Brazil, some of the labour and social issues were resolved by providing the affected parties with share in the newly privatised companies.

Introduce New Regulatory Structures

Privatisation takes the state out of the energy business but, on the other hand, increases its responsibility for setting the rules for the sector. In many countries the state is now moving in the sector—reflecting the concerns of the public—and leaving to markets the investment, operating and financial responsibilities.

As their economies have developed, most countries have tended to reply on specialised political process for regulating the energy sector and the utilities in particular. The best known of these is the US system where independent, quasi-judicial bodies have been established to set and enforce the rules. Most other developed countries have similar mechanisms.

The important point is the openness and inclusiveness of the process. Without this level of disaggregation of the political process, the growing complexities of regulating the energy sector can overwhelm the existing political institutions.

Restructure Prices and Contain Subsidies

Subsidies in the energy sector are difficult to control and seldom reach the intended target. They are inclined to end up in the hands of the politically rather than the economically or socially deserving.

They introduce large inefficiencies in fuel choices and fuel uses. Free electricity in rural areas in India, for example, has contributed to excessive and wasteful pumping of water.

Learn from the experience of Others

The pre-condition is clearly perceived : 'more private participation and market' wherever feasible and a new role for government. For this reform process, the lessons which emerged from the present Study are:

- reform seldom comes from within
- fast is better than slow
- successful reforms in any sector require a context of broader, social, political and economic change
- it is not easy to control or plan the reform process—once started, it develops a momentum of its own.

Learning from experience is a formidable challenge. Learning from experience gained elsewhere is an even greater challenge for governments and business, and facilitators such as the WEC.

If the Study demonstrates successfully the variety of options and their benefits in terms of finance and efficiency, it will have fulfilled its aim.

REPORT ON THE ACTIVITIES OF THE WEC ENERGY FACILITATION PROGRAMME : 1997-98

10

Dr. Rangaswamy Vedavalli
Manager, WEC Energy Facilitation Programme
London (U.K.)

Background

The WEC in November 1997, decided to expand its facilitator's role by instituting the Energy Facilitation Programme (EFP) to assist in getting things done to attract private investment in lesser developed countries' energy sectors of Africa including support to the African Energy Commission (AFREC) and South-Asia. The EFP was set up initially for one year (September 1997—August 1998) to operate separately from, but in parallel to the regional programmes.

The EFP objectives are to provide advice and recommended actions to government and industry to work towards rationalizing local energy operations, to examine changes in fiscal and market regulation to set up improved frameworks to attract private investment and to build up relevant and efficient institutions to mobilise local private finances.

Activities

In the short period , since the EFP has been launched in November, 1997 in Bangkok, several EFP activities have been carried out to achieve the EFP objectives both at the regional and national levels. In addition, concrete demand for over 25 EFP regional/national activities have been received from member committees in Africa and South Asia.

Regional EFP highlights include

- Organized tripartite meetings between WEC, UNECA, and the Organization of African Unity (OAU) to review OAU's report on AFREC. Provided advice and disseminated OLADE's experience to OAU in setting up of the African Energy Commission (AFREC) to promote African energy projects to investors.

- Participated and made presentations on addressing the challenges to attract private investment to African energy sector at the UNECA Conference of African Energy Ministers in Durban, South Africa in November 1997; established successful contacts with African Ministers, strengthened the partnership with the OAU and UNECA which resulted in recognition of the WEC's facilitator's role in Africa.

- Provide advice and facilitated the Conference on Financing Energy in South Asia to address national and regional energy financing issues and to apply actions to the recommendations of the WEC Study on Energy Financing. The Indian Member Committee organized the Çonference by raising the resources locally.

National EFP Activities

At the national level the EFP focused on providing advice on required institutional, policy and investment conditions to attract private investment; to reform and build institutions; and to formulate energy sector planning and policies including natural gas utilization and renewable energy. Specific EFP highlights include.

- *South Africa* : Dialogue on energy issues in South Africa including energy policy formulation and review of regulatory practices and bringing a perspective of experiences in the energy industry and their relevance to the local energy issues. Based on the dialogue with different stakeholders in the energy industry, several areas for facilitation both at the national and sub-regional level have been identified.

- *Ethiopia* : Initial advice on privatization of Calub gas project

- *Nepal* : Review of Nepal's institutional/regulatory framework to attract private investment in hydro-power development ; dissemination of regional and cross-regional experience in South Asia and Africa to attract private investment in hydro power development and their relevance to Nepal on specific issues licence period, formulation of power purchase agreements and reducing risks and profit-sharing between private developers and the government.

- *Sri Lanka* : Review of Sri Lanka's efforts in planning for sustainable development utilization of natural gas in power generation and facilitation of regional meeting on Biomass as a source of sustainable en :rgy.

- *India* : Facilitated applying actions to the recommendations of the WEC study on "Financing the Global Sector : the Task Ahead" to the national and regional energy financing issues in South Asia by supporting and providing advice to organize the Energy Financing Conference in Delhi (India), and setting up of the regional workgroup to address energy issues and apply actions.

Assessment

The above record speaks for itself to the success of the EFP programme launched in November 1997. Even within a short period of less than a year, the EFP has brought obvious benefits :

- added value to WEC membership in Africa and South Asia, raised the image of WEC both nationally and regionally and generated interest among non-members to become members of the WEC.
- emerged as a basic tool to develop bottom-up approach to address lesser developed countries energy issues in the context of globalization. The WEC is seen in Africa and South Asia as a responsive partner in carrying out the dialogue on energy issues within the energy industry both public and private.
- diversed the structure of member committees to include different shareholders in the energy industry Member committees have further enhanced their cooperation among themselves and across the regions.
- forged an effective link between national and regional energy issues and underscored the need for effective actions to strengthen regional cooperation in the emerging interdependent energy markets.
- strengthened effective partnership between WEC and sub- regional/regional and other multi-lateral organizations dealing with sustainable energy, environment and regional cooperation.
- facilitated crossing the gap between theory and practice in recognizing the relevance of recommendations of WEC Congress, Regional forums and studies to address local energy sector and region-specific energy issues. The EFP thereby is emerging as a vehicle to facilitate the process of "Thinking Globally and Acting Locally".

The Way Forward

What is required now is to consolidate on the benefits and move forward to start getting things by appropriate selection of EFP activities. Given that the process of energy transition is at the beginning end of the scale in Africa and South Asia, there is a long way ahead for countries in these regions to achieve the WEC objective of supply and use of sustainable energy for the greatest benefit of all. It must be acknowledged that it is the responsibility of countries to get things done to develop their energy sectors by creating enabling business environment, and much needs to be done by the countries themselves. Nonetheless, WEC can play an effective catalytic role as facilitator both at the national and regional level to promote private investment in energy development in developing countries.

OVERALL ENERGY SCENARIO : A CII PERSPECTIVE

11

C. Dasgupta
Member-CII Energy Committee,
Dy. Managing Director, CESC Ltd., Calcutta

PRECONDITIONS OF FINANCING

State Clearances

Power sector being under State control, all the primary clearances for putting up a power project are to be obtained from the State Government. Most of the cases, State Governments identify a project and select the promoter through MOU or competitive bidding. The statutory clearances from the State Governments departments are:

- Section 29 clearance
- SPCB clearance
- Section 44 clearance
- State Water Authority/Irrigation Department clearance
- Clearance from Land Revenue Department

Clearance from CEA and MOE&F

The technical and commercial evaluation of the project proposal is done by CEA when the Detailed Project Report is submitted by the Promoter after obtaining the above mentioned State Government clearances and the fuel supply is tied up. Necessary amendments, as recommended by CEA on various technical and financial aspects of the proposal are taken care before issue of techno-commercial clearance of CEA.

While reviewing the proposal, CEA also takes cognizance of the terms and conditions of the Power Purchase Agreement and a firm Fuel Supply Agreement.

MOE&F reviews the Environment Impact Assessment Study Report and Environment Management Plan prepared on the project proposal and issue the environmental clearance for the project, when satisfied that the pollution from the project will be within the applicable norm and adequate measures have been proposed for that including the provisions for management of a disaster.

Clearance of financing from DEA

The details of the project finances including debt equity ratio, sources of debt finance with moratorium period and interest rates, sources of equity, etc. require Department of Economic Affairs' clearance before going in for sourcing the funds. Obviously, all the statutory and non-statutory clearances are to be obtained and all the project contracts/ agreements are to be in place before obtaining the clearance on financing of the project from DEA.

Risk Analysis and Sensitivity

Financing a power project by the Bank are exposed to various risks, which a Banker will obviously look for sharing by the Promoter or the Contractors. Beyond that the Banker will like to cover the other risks by way of suitable insurance coverages. The risk coverages by different agencies involved in a project can be put as :

Power Purchaser : Delay in construction of transmission line; Force majeure delay beyond insured level; Demand risk and grid indiscipline; Interest/currency risk; Political force majeure; Construction Power delay; Country risk-Change in law/tax.

Promoter : Construction force majeure delay; Operating failure beyond the cap in EPC; Operating cost increase beyond tariff; Fuel transportation risk.

EPC Contractor : Delay in commisioning; Performance shortfall with a cap

O&M Contractor : Operating failure;
Increased cost of spares and manpower

Fuel Supplier : Default in fuel supply

Insurer : Force majeure delays; Operating failure (damage)

It may be noted here that apart from the risks mentioned above, finance from the Banker is subject to other risks like technological risks, risk of non-payment by the Purchaser, etc. With opening up the power sector to private entrepreneurs drive to achieve ever higher levels of efficiency makes the Bankers feel that they are approaching the leading edge of technology where generating plant would be designed to deliver previously unknown levels of efficiency as well as availability.

Once the construction and operation related risks are shared by the Purchaser, Promoter and the Contractors, the next important thing for the Bankers is to ensure that as far as can be predicted cash flow from the project will be sufficient to cover the operating cost and expected debt service requirements with reasonable margin of safety. A base case is prepared and individual input factors are varied singly or in combination in a range of sensitivity cases, to determine which factors are likely to reduce or eliminate the margin of safety for the Banker.

Cover Ratios like Annual Debt Service Cover Ratio (ADSCR) or Loan Life Cover Ratio (LLCR) provide the Banker with a standard yardstick which can be applied across a range of sensitivity cases, to test the impact of changing particular input variables. These ratios are the main tool of economic comparisons where Bankers determine the likely adequacy of cash flow from a debt service point of view, each year during a project's life. ADSCR is calculated by dividing the available cash flow, after operating costs and taxes in any given year, by the sum of interest and principal repayments projected in that year. Traditionally Bankers will look for a figure to the tune of 1.3 in any given year in the project's life.

LEVERAGES AVAILABLE FOR FINANCING

Debt Equity : 80% ; 100% Foreign Equity

As per the guideline from Ministry of Finance, debt equity ratio for a power project can be as high as 4 : 1. It is, however, left to be seen that whether such a debt equity ratio would be acceptable to the financiers of power projects in India with associated risk elements.

Ministry of Finance, in the recent past, has declared 100% foreign equity to be acceptable for power projects.

CONCERNS OF FINANCING

High Interest for Re-Loan and Volatility of FOREX Rate

Rupee Loan from IFIs with high rate of interest in the order of 18-19% is obviously a deterrent for funding more Re Loan besides its availability. The variation of the rate of foreign exchange is also beyond anybody's guess that how much the actual impact on the project cost as well as the tariff will be at the end of the day. The premium required for availing the foreign loan varies depending on the risk elements and credit rating of both the Purchaser and the Promoter. The funding of increased project cost due to devaluation of Rupee is also a problem during execution of the project.

Problems Related to SEBs

Due to long term nature of financing, the poor credit profile of the SEBs and the imminent reforms and liberalisation of the sector, investment in the power sector carries a high risk at the present time. As a result lenders are asking for suitable credit enhancements such as escrow account mechanisms backed by sovereign/State guarantee and more equitable fuel supply/transportation agreements before financing the projects. The States should develop a long term perspective plan on the reform process so that the projects can be structured to be viable even in the post reform scenario.

SEBs have so far catered to the social development as an extended arm of the government but lack a commercial orientation. High system losses, subsidised tariff to agriculture/domestic sectors coupled with political interference in policy decision as well as day-to-day operation has made the SEBs chronically sick, resulting in very poor credit rating Areas of operation covering generation, transmission and distribution being widespread, the size and structure of the SEBs have become unmanageable.

The legislative and market reforms currently underway in India are part of a seven years economic programme supported by the World Bank which has been designed to promote competition and private sector investment in the country. The reforms include the division or unbundling of power utilities into generation, transmission and distribution units.

Inadequate and Disproportionate Transmission and Distribution System

In India, power generation being predominantly coal based, most of the mega thermal power projects are located near the mines. The HV transmission system being inadequate, the power situation varies from State to State and even within the State itself because of inadequate distribution system. For example, while eastern region is surplus in power, the northern as well as southern region are running with huge deficit of power. Again in spite of eastern region having excess power, lack of adequate distribution system results in high system losses as well as lower PLF of the generating stations.

The wide variation in voltage and frequency in most State grids create problems for private power producers for smoothly synchronising and wheeling power to their purchaseRs. There is already indiscipline in sharing Central sector power by the States. Some overdraw power during their peak hours dipping down frequency below 48 cycles and dump power during off peak hours boosting frequency above 51 cycles. The Power Grid Corporation of India could punish such indisciplined operation by arranging automatic summation of power drawals by States and tripping by carrier transfer the radial loads of defaulting States when over drawal persists for fixed time and simultaneously frequency also is below 49 cycles. So also dumping can be punished by tripping specified smaller generators of defaulters when dumping occurs along with rise of frequency above 50.5 cycles. But there is neither legislative compulsion nor political will to act. This cannot continue with private sector coming in.

Also the near bankruptcy of most SEBs prevent investing adequately on needed transmission systems, substations and distribution and on needed reactive compensation. Proper power distribution without excessive voltage drop would be very difficult by the private sector. Due to lack of political will, they do not also undertake even regulatory measures to flatten load demand curve to improve voltage levels.

MODE OF FINANCING AVAILABLE

Indian Financial Institutions

Among the Indian Financial Institutions, ICICI, IFCI and IDBI are coming in big way to finance the power projects in India. As the development of the infrastructure sector is increasingly being undertaken by the private sector, they anticipate a reallocation of the relative sectoral investments away from traditional industry into the infrastructure sector and the power sector in particular. Although the role of IFIs has been purely the provision of direct credit, the market requirements is likely to change as the financial markets develop and deepen. ICICI is redefining its role to being a provider of comprehensive financial solutions for all kinds of projects.

RECOMMENDATIONS

Innovative Structuring of Finance

The structuring of finance for a project should be tailor made and innovations are required for structuring the finance for a particular project depending on the type of project, the credit rating of Purchaser as well as the Promoter, the terms and conditions of the Power Purchase Agreement/Fuel Supply Agreement/Fuel Transportation Agreement/EPC Contract/Project Management Contract/O&M Contract.

The financing for Balagarh Power Project being developed by Balagarh Power Company Limited where CESC Limited is the major promoter as well as the sole purchaser of energy, is tabulated below :

Equity	*% Holding*	*Debt*	*% Holding*
CESC	35 .	IFC 'A'	6
ICICI	5	ADB 'A'	6
SIEMENS	10	IFC/ADB 'B'	40
IFC	7	IFI LOANS	10
ADB	7	HERMES	38
AIDEC	6		
BHP	15		
AIG	15		

Retain Cost Plus Approach

On opening of the power sector to private entrepreneurs, two part tariff method has ben made applicable where predetermined rate of return for the investment is ensured after meeting the operating expenses and debt service requirements. Since the progress so far is not encouraging, some section of the industry is advocating for single part tariff with the logic of improvement in efficiency and transparency. But the real problems lies with the volatility of government policies along with host of other problems. It does not sound practical that private power will see reality when single part tariff is implemented without any assured return. It would rather be fair if performance related incentive or penalty are imposed over and above the assured rate of return.

Effective Implementation of Regulatory Commission

The setting up of the electricity regulatory commissions is a significant step in the reform process. Just as the setting up of the TRAI was a catalyst in the financing of the telecom sector by international investors, the creation of the electricity regulatory commission will be viewed as a watershed in the Indian power sector.

The commission's principal role will be to examine issues related to power tariffs and will be the highest authority to regulate power projects within the country. Some regions have independently made moves to install regulatory agencies as part of a wider policy to privatize and restructure the State utilities. The States of Orissa and Haryana have already passed legislation to install regulatory bodies, while the State of Andhra Pradesh suffered widespread civil unrest when it passed its own electricity reform bill.

States to Declare Power Policy

Each State should immediately declare power policy along with the long term perspective plan on the reform process. This will enable the project promoters to decide or prioritise the projects for execution.

It is also recommended that the projects will be opened to competitive bidding under a proposed "shell-company" package. The "shell-companies" will obtain approval for the power project and also many of the technical and environmental clearances needed to construct a power facility at a pré-arranged site. Projects would then require "no further clearance", only financing. The projects would be located at coal pit heads or on the coast to minimize fuel transportation cost.

Private sector Power projects likely to come in Ninth Plan

Name	*State*	*Capacity (MW)*
Thermal Power		
Mangrol Lignite	Gujarat	250
Sabarmati (AEC)	Gujarat	120
Bina	M.P.	578
Budge Budge	W.B.	500
Balagarh TPS	W.B.	500
Torangallu	Karnataka	260
Korba West	M.P.	120
Tuticorin	Tamil Nadu	525
Zero Unit Neyveli	Tamil Nadu	250
North Madras II	Tamil Nadu	1050
Rosa TPS	U.P.	587
Vizag TPS	A.P.	1040
Mangalore TPS	Karnataka	1000
Bhadrabati TPS	Maharashtra	1082
Korba East (Daewoo)	M.P.	1000

Name	*State*	*Capacity (MW)*
Bhilai	M.P.	574
Ib Valley 3 and 4	Orissa	420
Hirma	Orissa	3960
Barsingsar	Rajasthan	250
Bakreswar	W.B.	420
Gouripur	W.B.	150
Ramagundam	A.P.	520
Hydel Power Projects		
Maheswar	M.P.	400
Bhivpuri PSS	Maharashtra	90
Baspa	H.P.	300
Gas Projects		
Bhaskhandi	Assam	16
Adamtilla CCGT	Assam	9
Godavari CCGT	A.P.	116
Jegru CCGT N	A.P.	76
Paguthan CCGT	Gujarat	655
GIPCL Extn.	Gujarat	160
Hazira WHRU	Gujarat	185
Dabhol CCGT I	Maharashtra	740
PPN CCGT	Tamil Nadu	330
Bawna Ph-I	Delhi	421
Dabhol CCGT II	Maharashtra	2450

PART-IV
PANEL DISCUSSION

PANEL DISCUSSION ON IMPACT OF FUEL SUPPLY ON FINANCING

12

1) D.V. Kapur
Chairman, Reliance Power

Panel Chairman

Issues in raising financial resources have become more complex primarily due to scarcity of public funds not only in India, but all over the developing world. Independent Power Producers became relevant. IPPs were not promoted just because others were not able to do the work, but there was a need for supplementary development of private sector to fulfil the energy targets. The need arose out of inadequate financial resources with the developing countries, or in the developing countries. And, therefore, the concept of IPPs was developed in which neither the government invests in total, nor does any private developer invests entire. All the players depend on the lender, who has adequate financial resources, somewhere in the world. And this lender is 'the lord' and 'the master' at whose command everything has to be done. The general impression is that success takes place because that 'master' in turn puts stringent conditions for risk coverage and insists on project completion on time. All the commitments are not made on today's cost. If one is not able to execute a project in time and complete commitment in approved costs, the risk is his and not that of the financier. India is not used to that type of practice.

Total planning process in India is based on current costs. To tide over the inefficiencies and poor management some contingencies are provided. And that is the end of costing. Performance is analysed it will be noticed that methodologies adopted are such that these will be delays and cost over runs. I have spent enough of my life in the public sector and I am not critical of public sector but certainly of the administered methodologies. The IPPs have taken care of such methodologies as the lender does not agree for the same.

The lender ensures all this through specific contracts, with conditions and conditionalities to cover the risk. In that context it is not only agreement for sale of power or recovery of dues, the lender also wants bankable contracts that are forceable contracts. These contracts cover supply as well transportation of fuels.

The Chapter on Panel Discussion has been written on the basis of transcription of presentations. As each one of the Panelists spoke on the same topic, their presentations have been compiled under the Panel Topic. The Report starts with the presentation from the Chairman of the Panel who identified issues for discussion.

India and other developing countries are not used to this kind of contract procedures and therefore have teething troubles. That opens up new areas of new risks, new restrictions and new approaches in management. Further, in some other developing countries, like India, the ownership of the utilities; the fuel suppliers; and the fuel transporters all lies with the government—Central or State. It was for one government body or entity giving a commitment to another government entity. And the results are all very well known. Coal India as a fuel supplier blames utilities for non payment of millions of rupees, and NTPC as a power producer blames State electricity boards for non-payment. All of them have realised the need for enforceable commercial agreements. On one case go to the court of law because of lack of an enforceable legal document for fuel supply, transportation of fuel or power purchase. If commercial practices have to take over the practice then these has to a general agreement. This will be in the long term interest of the sector.

Specific Points for Discussion

Certain specific points for discussing will be as follows :

(i) *Legally enforceable documents for lending or financing* have not been clearly understood in many energy sectoRs. That is the first subject to be debated, in entire, including the fuel supply.

(ii) The next point to mention in this regard is the *capital cost of a power projects* which depend on the type of fuel chosen. For gas and liquid fuels supply costs are more or less similar. For coal, it is probably 30 to 40 percent higher.

(iii) Next important issue is the *financing needs for development of fuel itself.* When scanned in the international scenario both oil and coal are freely tradeable commodities. In most of the countries, they are under the control of large well managed companies under very very rigid, competitive environment. There is lot of politics, particularly in the oil sector. But that is a separate issue. But, one thing which is acceptable is that most of these companies, whether they are Australian coal companies or international oil companies, they also have huge amount of resources. Sometimes they exploit the strength of those resources. The lack of financial resources in the developing world has introduced the concept of "high regulation" and "fuel linkages". But that linkage and regulation also created its own problems. Respective companies not get used to competitive pressures. Privatisation and open competition are obvious answers and the governments in India and other developing countries should accept in principle and implement. Tremendous amount of additional finance is required in the Railways to meet the transportation needs of coal. By the year 2006, Railways shall have to dedicate more than three times the present capacity to coal sector. Before I mention the issues necessary for discussion let me say that I am personally not happy the way privatisation has been introduced in the power sector in India. The discussion in India is still in terms of public and private sector. Under the concept of privatisation, when a project is selected by competitively under open bidding, where every company from public sector or private sector should participate. Then, the contract should be offered to whoever meets the requirements, whoever can give the lowest cost. These types of issues are still raised and debated.

The functions of good management are doing things in time, doing more efficiently, raising productivity, achieving lower cost of production.

(iv) *Fuel processing* is another area for financing. It is not only a question of production of fuel. India has very high ash content coal. It amounts to transporting 40 to 45 percent ash. To what extent it should be washed and what technologies are available are related to finance. Similarly, in case of high sulphur the basic concern is howto restrict sulphur contents in coal. New technologies are there, new types boilers are there, possibility of IGCC technologies is coming up very fast. Somewhere in the world, high sulphur crude is available at a low price, which can be converted into gas and used. The issue is ultimate cost. It has to be more competitive. But, all these need additional finance.

(v) *Transportation and handling infrastructure for fuel* is yet another area for large financing. I mentioned about Railways. How much capacity they will require to expand. There is also inadequacy of pipelines, tankages, jetties, specialised wagons. All these require large amounts of financing. Government has again accepted in principle that financing of these areas including, ports should be done through privatisation. But there has been little action. It was not a question of 'private versus public', in their traditional sense. The question was non-availability of adequate money within the country. So whether it is going to Indian private company or a foreign private company, all of them are partneRs.

(vi) *The fuel is the major component of cost of power.* Ideally the lowest tariff should be the main criteria for selection of fuel and investment on the project. But developing countries have their own constraints primarily due to lack of foreign exchange balance. However, in India, inspite of the fact that coal is our fundamental source of which we have fairly large amount of deposits, which will last us for many years, there is a large part of the country, where Indian coal will not be able to reach at an economic cost. Start from Gujarat, Maharashtra, Karnataka, Kerala,

Tamil Nadu and Andhra. Lucky all these States have coastal line. On a long term basis there will all have to depend on imported fuels. So the locations to be selected should have the least inland or overseas freight. And, in all those cases, when the projects are put forward to competition today, owner whoever it may be—whether it is State government or Central government should not interfere with regard to fuel for a particular project. So also its capacity and location. The fuel should be selected on the lowest cost of generation, whether the contractor uses fuel like oil-emulsion, LNG or imported coal. In any case it has to be imported fuel. The concern which requires attention is tariff. Liquid fuel will have lower capital cost and solid fuel will have higher capital cost. An example is that LNG is a growing area. But in case of LNG simultaneous investment for a 2500 MW power project would require 2 billion dollars of power station, another half a billion dollars of LNG terminal in India, almost another 2 billion dollar required to develop the gas field and liquidation facilities in the gas producing country, which also requires financing by the foreign lendeRs. So the amount of financing in that case becomes so large but again it has to be done. It is happening in the world.

These are certain issues that the Panel should address

PANELISTS

2) Devi Dayal,

Addl. Secretary Ministry of Petroleum and Natural Gas.

The presentation covers supply of liquid fuels in India. Every project requires four basic elements :

i) Finance;

ii) Fuel finance;

iii) Technology and the equipment; and

iv) Marketing

All these are covered under essential arrangeme' t, named fuel supply. To achieve any financial assistance it is necessary to have assured supply arrangement of fuel.

The present installed generating capacity is over 86,000 MW. The energy and peak shortages are reported to be 7 to 12 percent. So the present shortage in India for power is in the range of 10-15,000 MW. This is likely to continue for a few more years unless remedial measures are taken up. To avoid crisis the government considered to bridge this gap. The main source of power supply or power generation in this country is coal and hydel power.

These are the types of issues which need to be debated in the country and a consensus needs to be arrived. And the country of course is taking steps for promoting power stations on these two sources. However, other alternative fuel the natural gas is getting to be a popular fuel for power generation throughout the world. We do not have sufficient gas production in the country. Our present supply of natural gas in the country is only around 57-58 mcum per day against the much higher demand, and this gap will continue to grow in the coming yeaRs. Thus, there is need for import of gas in some form or the other. Earlier, India considered importing natural gas through pipeline from Oman and Iran. However, those projects are not finalised so far because of some technical and other problems.

The immediate solution is the import of natural gas in the form of Liquified Natural Gas, LNG. LNG trade in the world at present is 70-80 mt and major importers of LNG are Japan, Korea, China and USA. There are number of suppliers of LNG in the world and this business is generating in all the major companies of the world and they are very keen in supplying LNG in our territories. Some foreign companies have got approval for import of LNG in the country. The Government of India has taken initiative for importing LNG in the country to meet the energy requirement of the country. Leading oil companies, ONGC, BHEL, IOC and Bharat Petroleum Corporation import LNG. These companies are making good progress. They have already selected two locations for import of gas, one at Kochi and the other at Dahej. And they are trying to tie-up customers also. It is hoped that in the next four years, LNG will be made available in India. Depending on the success of this initiative the private companies will be interested in putting up LNG terminals in the country. However, this involves a minimum of four years, and probably it may take longer time for other countries to put up the LNG import facilities in the country.

So, the immediate solution is found in the form of liquid fuels. The availability of liquid fuels, the cost of generation of power on the liquid fuels and other considerations influence such a decision. It was decided that about 8-10,000 MW capacity should be promoted on liquid fuels. In December 1996 the government decided to allow this capacity. The decision was then extended to 12,000 MW to different States, depending upon the present installed capacity required by the end of the Ninth Plan. Ministry of Petroleum and Natural Gas, on the recommendations of the State governments and the Ministry of Power has already made allocations during the last one and a half years.

Unfortunately the progress of IPPs in this regard has not been encouraging. We made allocations of neptha, fuel oils, and NLHS to 117 parties for putting up power projects based on these fuels. Out of these, 94 parties have signed the fuel supply agreement, supplied bythe oil companies. Four more parties against earlier allocations have also signed the fuel supply agreements. Considering the fact that they are already making efforts with the State governments to tie-up the PPAs and also with the financiers to lend the finance they are given 6 months. This was not availed by them, so they could not sign the agreement. This time was further extended and now it is learnt that only 4 parties have achieved financial closure. But there are number of parties, that have reported to us that they have made substantial progress with their financieRs. It is hoped that few months from now many of them will be coming out with the power projects.

For the information of those who are interested and are applying and requesting for allocation of more naptha or neptha for new power projects. The information is that the government is considering the same. The High Powered Board considered this issue and now they have decided that this 12,000 MW capacity limit will apply only for Neptha. Earlier, there was an allocation of fuel oil and LHSL also within the capacity of 12,000 MW. Now it has been decided by the government that there will be no restriction on putting up power plants based on liquid fuel of any capacity. So the States are free to promote power plants on fuel oil and LHSL to any extent. Similarly it has also been decided that there will be no restriction on the use of gas and oil emulsion for the generation of power. So this is also now free. State governments can use their discretion either to change the allocations or select new parties, or both. Permission has also been given to the State governments to readjust among the existing parties or to cancel their allocations or allottee in favour of a new party, on the basis of the competitive bidding State governments are allowed time upto 31st October, 1998. However, if they want to change the allocation, want to select new parties, then of course, 6 months time will be given to them for making selection, competitive bidding and further six months time to the new parties for achieving financial closure. So far as the terms of fuel supply agreements are concerned, there had been some controversy in the past. We had circulated a fuel supply agreement among the IPPs and all others, including the financial institutions, State governments, SEBs etc. The Ministry of Power has also hired consultants to advise us on the drafting of the Fuel Supply Agreements. Based on all that Fuel Supply Agreement was finalised for the month of September and circulated to the IPPs.

Mention may be made of a few points. first of all whether the oil companies are ready to pay liquidated damages for the failure to supply the fuel. It has been agreed by the oil companies and incorporated in the fuel supply agreement that they will pay the requidated damages to the extent of 100% depreciation; 100% of interest paid; 50% of the office and maintenance charges and 100% of Insurance paid during the course of the year for the shortfall in the supply. Therefore all the companies have stipulated that IPPs will have to pay a guarantee charge. This guarantee charge was fixed at 5% for 68.5% committed supply. However, all companies have made some concession in this regard, as the companies will be required to put up new infrastructure in the form of jetties, tanks and other facilities for the despatch and supply etc. Therefore, naturally they put infrastructure charge also to ensure that this facility will be utilised. They have also asked for commitment charge. Earlier this commitment charge was supposed to be realised in the form of cash as well as bank guarantee but now some companies have decided to accept the charge in the form of bank guarantee only. The lender is concerned about the capacity of the transporter and the oil companies to supply the fuel in time. So it has been tied up. An arrangements has been worked out with railways. Some extra cost should be paid to the railways to ensure that the liquid fuel supply is transported in time. It is also ensured that there must be a minimum five days stock to avoid the shortfall in the production. All the companies will be responsible for the liquidated damages.

By and large this Fuel Supply Agreement has been found fully acceptable and there is good response. An offers has been made to the IPPs and some of the foreign consultants for providing a draft for the Better Fuel Supply Agreement, so that Fuel Supply Agreement serves fully for the purpose of developing power on liquid fuels.

The financiers as well as developers seem to be unnecessarily apprehensive before investing money in the power sector in the country. The situation is not so bad. Companies like G.V.K. Power Corporation and Spectrum Power Corporation are already running on natural gas and liquefied gas supplied by the oil companies inspite of the fact that these companies had no formal agreement in the supply of the liquid fuel. But there is no problem. The Indian oil companies are competent and they have never failed in their commitments.

The Central Government has already come out with an act for compelling the State governments and the Central government for setting up Regulatory Commissions for Tariff. It also declared that if there is any under recovery on account of any supply at lower cost to any section of the society, the State government has to compensate out of its budget. Indian companies have been fully geared up to deal with the liquid fuel supply.

3) Tony Concannon
Leading Member of Hinduja National Power Coal Supply Negotiating Team

Thank you for inviting me to give a short presentation title "Impact of the Fuel Supply on Financing of Power Projects" to this forum today.

By way of an introduction I have worked in India for over 4 years on a number of power projects but in primarily the Vizag 1040 MW coal-fired Project in Andhra Pradesh. This project uses washed indigenous coal; is currently finalising its financing and the deal with Mahanadi Coalfields Limited based in Orissa is virtually complete.

Independent Power Produce projects (or IPPs) are complex deals and a number of issues are inter-related. I do not intend to address all of these but I do intend to give a flavour of how the fuel supply chain fits into the overall framework of an IPP and in particular what the Lenders usually focus on.

BASICS : GENERIC TO IPPs

Demonstrate fuel in; electricity out

IPP finance depends on documentation*

I have set out two basic points.

The first is a macro view taken by Lenders (and indeed most Sponsors) when they begin to appraise a Project can the fuel get in and can the IPP get the electricity out? although both fundamental, and you would think they would be sorted out from day one, there are examples around the World where these basics have been lost in the detail and the Project has fallen apart prior to securing finance or has been in distress from day one.

The second, is a reminder that non-recourse IPP's must be able to demonstrate to lenders that it can exist as a stand-alone power business.

Interfaces with all external parties are usually documented using agreements (e.g. EPC Contractor, power purchaser, stake-holders and of course the fuel supplier).

LENDER REQUIREMENTS

- *Robust fuel linkage msut be in place*
  - — *Credible suppliers involved*
  - — *Customer needs are addressed*
  - — *Environmental and social matters*
  - — *Alternative supply and transportation*
- *Agreements signed: incentives and penalties*

Whether the Project is in Europe, South America or India the requirements generally remain the same from the Lenders regarding fuel supply. The Project must be able to clearly demonstrate that it has a robust and credible fuel supply chain. **Contrary to popular belief, lenders do not simply rely on contractual recourse to get comfortable, they must also be satisfied that the entire fuel supply chain will ultimately work which involves appraising the players involved, their experience, and what contingency plans are in place including fuel storage and alternative supplies.**

Lenders are also becoming increasingly concerned about environmental and social matters (e.g. rehabilitation of people) both upstream (fuel source), and downstream (power evacuation) of the IPP.

The Lenders will carefully review if the needs of the customer are being met in terms of fuel quality, quantity and programme. Quality and quantity because the power station is only an energy converter: if the heat content of the fuel delivered to the Station reduces then the electrical output will ultimately reduce. Programme, in order to meet the real-time nature of electrical power supply, and the mandatory outages that complex power plant require.

Now I would like to turn the attention to IPP's in India. Although, there are IPP's using numerous fuel types such as Naptha, gas and fuel-oil, I am going to focus on indigenous coal. I do not intend to go through the long list of statistics surrounding India's significant coal (and rail) industry, suffice to say that coal will undoubtedly continue to be the primary fuel source for power plant in India for many years to come.

PRACTICE IN INDIA TODAY

- *Over 180 mm tonnes of coal transported per annum in India*
- *No record of power being lost due to lack of coal*
- *All fuel risk borne by SEB's*
- *Not an Agreement in sight....*

Over 180 million tonnes of steam coal are mined and transported by rail throughout India every year. There is no record of power being lost due to lack of coal at Indian power stations.

The SEB's pay administered prices set by the GoI for the fuel supply chain (coal and rail), and in return assume the full fuel supply risk which is ultimately passed onto electricity consumers.

It all works relatively smoothly and not an agreement in sight !

So why after 6 years has not one large indigenously coal-fired IPP, which uses significant offshore funds, reached financial close in India?

The factors involved are numerous and complex ranging from protracted clearance processes to tariff subsidies, and I am sure you are all well versed it what they are—we have certainly had enough conferences to discuss the same What intend to do instead is share with you my experience with Lenders on a large coal-fired Indian power project namely the Vizag Project.

EXAMPLE IN INDIA

— *Eight months of due-diligence to date*

— *Over 1000 questions responded to*

— *Site visits by technical advisers*

— *Numerous road shows to lenders*

— *Objective : building confidence*

As I mentioned in the first slide Lenders require a clear relationship to exist with all external agencies including the fuel supplier and transporter. This relationship is documented in the form of a legally enforceable agreement. With a State-owned to State-owned arrangement there has been no need for such contracts, but with the advent of IPP's this has changed. With no precedent in place there was some nervousness by the State-owned entities to enter into such agreements but this is now being overcome.

One cannot isolate the fuel supply process from the issues surrounding the SEB's. Two main ones are :

1. SEB's (and State governments) are now reluctant to take fuel supply risk even though the entities involved are today supplying their own power stations with coal (at the SEB risk); and
2. If the SEB cannot pay the IPP then the IPP cannot pay the fuel supplier and transporter. The credit worthiness of the SEB is therefore important.

FUTURE PRACTICE

- *Emphasis on customer needs*

  — *Quantity, quality and programme*

  — *Heat = MWH, not Tonnes*

— *Environmental constraints*

- *Leaders become more comfortable*
- *Increased use of agreements*
- *Coal expansion, dependent on power expansion (and vice versa)*

Looking to the future, I believe the process will speed up dramatically.

There will be a continued reliance on indigenous coal suppliers and rail transporters to deliver coal to power plant. They are more than qualified to do the job.

Precedents will exist and this will enhance the perception of the Lender community.

Agreements for fuel supply and transport will become common-place, and the needs of the customer will ultimately prevail (in line will international practice).

In addition, the coal and rail industry is striving to expand and I see ambitious targets across the patch but these will only be achieved if the power systems to evacuate electricity develop at the same rate. We are back to the second slide again !

Thank you for listening and I am sure the conference will be a great success!

4) Sanjay Bhatnagar

Managing Director, Dabhol Power

Five years ago, when Dhabol entered power generation in the country and proposed the use of LNG. The Idea of new fuel met with a lot of resistance and people believed that would not be possible. The actual idea was to introduce a new fuel like LNG, whose suitability under local conditions was to be tested. Now enormous support exist for LNG in the country. This just shows that proper decision can bring around required changes.

If the enabling legislation comes in place, then nobody restricts introduction of LNG into the country. Dhabol is now close to people's needs. Construction of first LNG terminal will probably start only towards the end of the year. By the year 2001, we should have the first LNG supply into Dhabol. Dhabol is building to introduce new fuel into the country, because the legislation has allowed that.

The panel needs to discuss:

(i) First, the type of fuel and who should decide as to what fuel should be used, and how that will be sources; and

(ii) Second, the types of contracts.

It is very important to have an 'enabling policy'. If the interest is in power development it will be better to discuss the price of power and leave the rest to the developers, it really works well if that is done. It is difficult to guess at what is the best fuel solution at any given place. For instance Dhabol was envisaged originally by the Maharashtra State Electricity Board to be a coal fired power plant. There were a lot of environmental concerns. Building Dhabol, people who did not understand at that point of time the benefits of a natural gas power plant were proposed to building power plant till they found what the truth was and benefits were.

Now think about coal fired power plant in Konkan area, at the same location the environmental outcry have been huge. A coal fired power plant at that very location would have produced 10,000 tonnes of ash every single day. Where would you dispose of that ash and what will you do with that? How would you reduce emission? What kind of equipment you would use? All those issues would have been really unwanted, and one came and suggested LNG became a possibility. Today, you can finance and build and commission an LNG fueled power plant, probably cheaper than coal fired facility. Some may say it is absolutely not possible. Our number shows it is possible and it would have happened. The choice of the fuel should be left to the developer.

But, there should also be a choice and the ability to diversify the fuel choices so that other fuels find a place. Liquid fuels were introduced as an interim measure as they allowed power stations put in a short time. India have a large need of power, not only in the cities, but also in the rural areas. So, how does one reconcile with a slightly higher cost of liquid fuel for power generation. Liquid based facility be built in the areas where one can switch over to gas. Dhabol Power is building domestic gas resource in the country; and LNG stations are planned all around the country. International gas pipeline may also be available from the east and also from the west.

A lot of people are concerned about the fuel price in India. Keep in mind that other countries have built considerable infrastructure within the country and paid for it, and are now reaping the benefits. A power producer, in India has to build the infrastructure and finance the same. In other words, 'pay to date for what is needed tomorrow'. At the start of Dhabol project, one of the things quickly realised was that a large project could be undertaken only if things were under control. So we planned Dhabol a project which did three things:

(i) it produced power,

(ii) it did own fuel supply, and

(iii) it built infrastructure for fuel supply.

So Dhabol is building own jetty system and mooring system. Dhabol Power can now take the fuel risk. The fuel supply is a natural choice of the person, who is best able to manage the risk. In Dhabol what was guaranteed was the availability to the SEBs. That is necessary for financing these projects and that will happen as soon as competition in fuel supply is promoted within the country. If there is mandate that power projects need fuel source from only one source and even the agreement is pre-determined, it really restricts the flexibility of IPPs and the lenders to fashion the fuel supply arrangements. The IPPs do not have freedom to choose. They should be able to choose in accordance with what the market demands. If an IPP can produce power using LNG facility cheaper than somebody else using coal, surely the market allow him the right to use that. If another IPP can produce power cheaper using coal, that should be allowed. This is the beauty of the market. In the market there is competition and competition always brings prices and choices to the customers.

Last comment about the various types of fuels bear customs duty and taxes. There should be rationalisation of fuel prices and tariffs. The projects should be closed today so that in three years the country will have surplus power.

IPPs are mostly impatient. They are not very comfortable waiting; and I think especially because the wait is not only a matter of time but also money. Considerable amount of investment made in power projects, most of the IPPs expect return of their money. IPPs deserve sympathy.

5) N.N. Gautam,

Advisor (Projects), Ministry of Coal, Government of India

Coal is the mainstay of power generation in India, producing about 55,000 MW. This is likely to go up to about 68,000 MW by the end of Ninth Five Year Plan (2002). Future projections for coal production are likely to be about 412 mt by the year 2002; 653 mt by the year 2007; and 559 mt by the year 2012. And, in this the share of coal for power generation would be about 267 mt; 447 mt and 559 mt respectively. The gaps between the production and demand in the year 2002 will be 54 mt; in 2007 the gap will be 235 mt and in 2011 the gap will be 263 mt. The gaps are not projected for want of coal reserve, but because of dearth of money the investment required in coal mining sector to produce coal. That is the reason the coal sector requires funding from private parties.

The main issue that can be seen is that when India looks for such finances, it has to look up to the lending institutions.

The main lending is to the power project, because it is a huge sector. To insulate from risks the lenders look toward, search for good contracts, because that is the only thing they can do as a reliable factor. There are 5 major contracts that they are looking for:

i) EPSE contract

ii) fuel supply contract;

iii) transport contract;

iv) finally if any company is involved in generation of power then PPA; and

v) BPL

Now, these are five major contracts, the lenders look to cover their risks. Now in case of coal being a fuel the fuel supply agreement is the one which has to be dealt in the panel. Drafting a good Fuel Supply Agreement is taking time. So far, the coal producers are the public sector undertakings; and the users were also government owned State electricity boards. Therefore, the legally enforceable agreement were not written and known. Suddenly, when these things are being discussed the lack of experience and apprehensions have surfaced. The dealing officers are taking time. The lending institutions, who are quite clear in their working are looking up for sound contracts and are laying down tough conditions. Whereas the one side conditions are very clear and tough; on the other side the lack of experience. This is taking time.

Fuel supply agreements have their own peculiarities. Whereas the lending terms are of 10-12 years only, the fuel supply agreements are being insisted upon for 30 years. Under the present working pattern it is difficult to predict the situation and get into an agreement of 30 years. The contractors have to project over 30 years and if they go wrong any where they will be irreparably harmed. It is difficult to understand as to why, fuel supply agreements are being insisted for 30 years. These should really terminate with the lending term period. This single contract has stopped progress in a number of projects. There are number of fuel supply agreements that have been initiated. But, then they are not progressing. It is projected that the IPPs are in a hurry. The government has allowed linkage to the IPPs. Based, on that linkages and that demand projection, the gap was worked out at the end of Ninth Five Year Plan (2002). Suddenly, the situation has changed, because the IPPs could not bring the plant on the time. There will be a surplus of 10 mt at the end of the Ninth Five Year Plan (2002)). Now, suddenly a reverse situation is coming. If this was also to happen in Tenth Five Year Plan (2002) would the coal companies be investing. The coal projects have a long gestation period. If coal has to be made available in the next 5 years, the investment decision has to be made now. So what investments were worked out in the Eighth Five Year Plan, thinking that there would be a gap today in the situation at the end of Ninth Five Year Plan, there is a surplus. And today in the coal sector there is a surplus already. The coal companies are not in a position to market, just because the programme and plan of the power sector have slipped.

Comment

Just a small comment on government, but you got to keep in mind there is international practice, because lenders sitting somewhere in the world, Tokyo or Wall Street he has not got a special love for this country. He is not bothered whether somebody fail to put the power sector goes fast or not, in time or not. His very limited interest is should I put my money in industrial project or should I put my money in power project or should I put my money in irrigation project. Even the Indian financial institutions are doing the same thing. So he wants every bit of documentation to be placed in before he finalised the deal. That is the point.

Sir, in this I do not know whether I gave the impression that I am saying, that the lender is not doing right, what he is doing.

6) S.K. Chowdhary,
Former Chairman, Coal India Limited

Fuel supply issue has to be carefully considered two points need to be understood and deliberated:

(i) Mutual interest has to be maintained. Wherever the power producers have taken interest in coal supplies, they have succeeded. NTPC is an example. They setup power stations near pit head, 40 to 60 kms of roadway and rail links were laid by them. They made their own merry-go-round. That means they made their own supplies and made sure that they have full interest in the kind of coal-coming. NTPC power stations are more successful.

(ii) Present status of lending and financing problems in coal industry need to be understood. They have got joint-ventures in coal mines. It will be worthwhile that power companies also invest in developing the coal mines; and the coal companies take due interest and sharing in the power companies. As the coal companies become strong they will have abiding interest in power companies.

SALIENT POINTS

P.K. Kanchan
Ministry of Coal

1. IPP is a new concept in India which has arisen out of compulsions of lack of investable funds. Appreciation of IPP as well as of lender has not been understood by the senior authorities due to total absence of practice of commercial agreements between government to government dealings. For faster implementation of power projects the requirements of IPP and lender has to be understood at the first place.

2. Indigenous coal suppliers and rail transporters will continue to be relied upon as they have the requisite capability. Coal sector in India needs tremendous investments for increasingly coal production. Capacity to transport coal by Indian Railways have also to triple by 2006-07.

3. Indian coal has high ash. It will be necessary to wash this coal to ash level to be determined by the location and other commercial aspects of plant.

4. In large part of the country e.g. Maharashtra, Gujarat, Tamil Nadu etc., Indian coal cannot be reached economically. These coast based locations can best be served by imported coal.

5. Developer of a power plant should have the liberty to decide upon the location of the power plant as well as the fuel it will use depending upon geographical location and other factors.

6. Selection of an IPP should be based upon the ultimate cost of generation of power and should not make any difference between an Indian private party or a foreign one or even between private or public.

7. Coal based route of power generation alone may not meet the growing demand of power in India. Alternative to this is in LNG, creation of facilities for import of which may take 4-5 years time. As an interim measure, liquid fuels eg. naptha, fuel oil and LSHS based short gestation power plant capacity has to be promoted.

8. Naptha based capacity can be up to 12,000 MW. Addl. capacity can be based on FO/LSHS using IGCC technology for satisfying environmental requirements.

Orimulsion and condensate can also be imported freely for power generation as a short term measure.

9. Model FSA for ensuring uninterrupted supply of liquid fuels to IPPs is in place and is eminently bankable and financial institutions are agreeable to fund the power projects. Both the IPP and the fuel supplier should utilise this opportunity to hasten the execution of power projects in India.

10. Back to back arrangement with railways are also in place for timely transportation of liquid fuel. This will also help in quick implementation of power projects.

11. Sincere efforts for the development of power system to evacuate electricity has to be made to avoid delays in implementation of power projects.

12. The facilities has been created for liquid fuels for the interim period should be such that can be changed for LNG latter.

13. IPPs should have been freedom to choose fuel right form now and should not be asked to wait for the year 2001 by which time the market forces are expected to take on as most of the power projects being taken up now will fructify in 2001.

14. Rationalisation of taxes, duties and other levies between various fuels is extremely necessary. Government should take up this issues urgently.

15. From deficit situation in the IX Plan, the coal is now in a surplus situation mainly because of the slippage of IPPs due to various reasons. The coal companies should not have any apprehension as this surplus situation is not likely to persist in future as initial problems in concluding FSAs are showing signs of overcoming.

PART-V
ASSESSING DOMESTIC CAPITAL MARKET

ACCESSING DOMESTIC CAPITAL MARKET

13

R.V. Shahi
CMD, BSES Ltd., Mumbai

If we start with the premise that Indian Electricity industry, dominated by SEB's, is in serious financial crisis and that any large capital investments in the sector may not be serviced, the whole exercise of funding solutions to the problem of power sector funding will be a non starter. Therefore it will be desirable to search solutions may be interim despite the poor financial health of the industry on the assumption that based on various initiatives that have been taken the situation would slowly improve. We need to recognise that the poor financial health of the sector has been there for several decades and yet it was possible to generate a growth rate in the sector in the order of 12.2% in 60's, 7.5% in 70's, 8.1% in 80's. It is only in the 90's, coinciding with the private power policy, that the growth rate declined to merely 4.3%. It needs to be restored to over 8% if the targeted growth rate of economy at 6% to 7% and industrial growth rate at over 10% are to be realised.

In this context, considering the past background, there is a need to have a change in approach to lending by financial institutions and also an approach of the Government to selectively backup such lending.

The size of the problem needs to be recognised and appreciated. In the next 10 years period for power generation projects alone we may need more than Rs. 350,000 crores. Besides transmission and distribution together with fuel sector would need commensurate investments. The debt markets have become unfavourable particularly in last one year. Interest rates have gone high and dollar rupee relationship highly volatile. As a result, external debt will be putting enormous burden debt on project cost and consequently on tariff. There is, therefore, need to tap the domestic sources to the maximum extent possible. It is in this background that following suggestions need to be considered.

i) The present possible of the GOI which restricts domestic debt to just 40% of project cost needs to be reviewed immediately. Depending on the nature of the project and its bankability, Indian financial institutions and other Indian Financial agencies could be allowed to provide debt to the extent they can and no such limit should be imposed.

ii) The sector expects a more favourable dispension from financial institution by facilitating faster financial closure, looking at risks, a little more kindly with higher degree of fragmentation and for advising the SEB's on one hand and IPP's on the other to arrive at a balanced package of terms and conditions. Indian domestic market also needs to consider a longer tenure of debt with necessary period of moratorium. It could be two to four years moratorium followed by a 10-year tenure.

iii) Power Financial Corporation (PFC) has emerged, of late, a major funding institution for power sector. They have taken a number of initiatives. Some of the suggestions for funding PFC to consider. Would be (a) organise and collect capital on long term both from India and abroad, cover foreign risk fluctuations and lend to developers in Rupee terms by charging additional interest rates to cover the foreign exchange risk. (b) In the changed scenario, PFC needs to shift to private sector in larger proportion than has been done so far. (c) It needs to adopt a more commercial orientation on interest rates which should be based on the credit rating of the borrowing agency rather than following a uniform rate. (d) A re-examination of strict terms and conditions to amend them at par with other agencies so as to make sure the terms more acceptable to the industry. (e) Tenure of debt needs to correspond to the need of the industry.

iv) Infrastructural Development finance Company (IDFC), which was created in accordance with the budget 1997, is organising on the basis of several important initiatives. IDFC could prove another major source for providing and facilitating capital for the sector. It could organise ECB, funds from agencies like World Bank, ADB, OCEF, etc. and lend to the sector on the basis of the Rupee terms, cover the risk of rupee fluctuation by suitably charging additional interest. IDFC could also evolve a suitable mechanism to provide suitable guarantees so that project developers could access debt market.

v) Recently in the budget 1998, industry has been allowed access to the fund available in the Provident Fund. Power sector being the core input, additional incentive could be provided so that the sector could access this fund. Besides, individual companies could also be allowed to access their own provident fund for power project development by themselves or by their sister companies.

vi) Infrastructure Bond, is a good instrument to access domestic capital market. However, the Government needs to clarify that total interest on income would be tax exempted. Besides there are a number of other clarifications on issues that have been raised by the industry that need to be clarified. Government may also consider allowing individuals to subscribe to infrastructure bond to encourage individual savings and generate fund for the power sector.

vii) Independent power producers arc invariably resorting to non-resource financing while organising funds for their projects. This obviously causes several hurdles in financial closing. While this option may continue, IPP's also should propose projects on resource financing basis which will provide much easier access and softer terms.

viii) Similarly parent company guarantees by the project developers could also facilitate faster financial closing and more comfortable terms of lending.

As stated in the beginning, on a long term basis the electricity industry has to enhance its credit worthiness by improving its financial health. This will call for reforms and restructuring of the industry and tariff rationalisation to generate surplus for growth.

There is no alternative to this if massive expansion of the industry is to be achieved. However merely because of the fact that the sector has the difficulty as mentioned, the funding exercise cannot be allowed to be a non starter. It is in this context that a few suggestions have been made in reference to accessing domestic capital market. All players in the field need to address the problem, perhaps with a larger degree of pragmatism.

FINANCING OF POWER PROJECTS-NTPC A CASE STUDY

14

C.P. Jain
Director (Finance), NTPC, New Delhi

GENESIS

Keeping in view the growing gap between demand for and supply of power in the country, the Government of India formed the National Thermal Power Corporation Ltd. in 1975 with a view to augment the efforts of the State sector in capacity addition and to plan, promote and organise integrated development of thermal power in the country. Initially, NTPC besides planning, engineering, constructing and operating power plants, was also undertaking the development of associated transmission network for evacuation of power from its stations. In 1992 the transmission business of the Corporation was transferred to the Powergrid Corporation of India Ltd., a company set up by the government of India for development and operation of transmission network in the country.

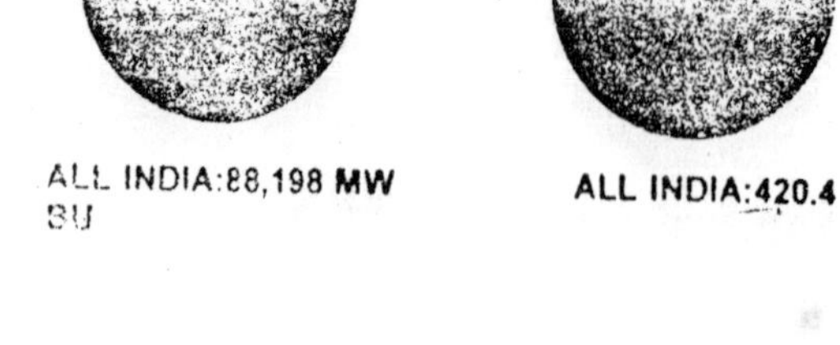

Fig. 14.1 : NTPC in Indian Power Sector

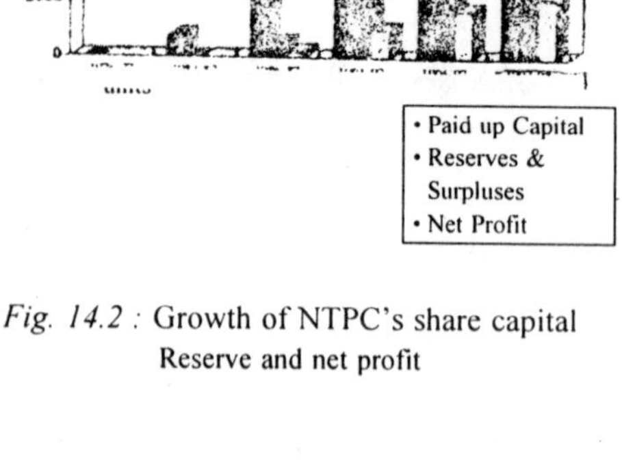

Fig. 14.2 : Growth of NTPC's share capital Reserve and net profit

GROWTH

The growth of NTPC since its inception, measured in terms of capacity addition, operational performance or financial performance, has been spectacular and perhaps, is the most important achievement of the Indian Power Sector in the last twenty-five years. As on 31st March '98 NTPC has an installed capacity of 16,795 MW (19% of total installed capacity in the country) at an investment of Rs. 27,471 crores. It generated over 106 billion units in 1997-98 (25.3%) out of total generation of 420 billion units in the country.

The growth of NTPC in financial terms has been no less impressive. The company made a profit of Rs. 42 crores in its very first year of operation in 1982 and since then, the profits have consistently shown an increasing trend. For the year 1997-98, NTPC profits are estimated to be Rs. 2,122.30 crores which is a 26% increase over its profit of Rs. 1679.43 crores for the previous year. Fig. 14.2 depicts the growth of Paid up share capital, accumulated reserves and net depicts the growth of Paid up share capital, accumulated reserves and net profit of NTPC since its inception.

NTPC'S OPERATIONS

Present activities of NTPC can be broadly classified into the following categories :

1. Generation of Power
2. Construction of Power Projects
3. Consultancy Services

Generation of Power

NTPC has an installed capacity of 16,795 MW at an approved cost of Rs. 25,402 crores. 13,620 MW out of the total capacity has coal as its primary fuel and the remaining 3,175 MW consists of gas/liquid fuel based units. The coal based stations consist of fourteen units of 500 MW and thirty units of 200/210 MW capacity. It may be seen that on an overall basis, the per MW cost of capacity for NTPC works out to only Rs. 1.51 crore which is a very low and makes it possible for the Company to supply power at the cheapest cost per KWh. Further, majority of NTPC coal based units are located at the coal pit-heads due to which the variable cost of generation is low for the NTPC units. Thus, the generation of power from NTPC units would work out to be an economical proposition in the event of Merit Order Operation of all the units in the Grid.

NTPC acquired the Unchahar (420 MW) station from Uttar Pradesh in Feb.1992 and the Talcher (460 MW) station from Orissa in June 1995 by adjustment of dues against energy sales. The performance of these stations at the time of take over was abysmally low; the PLF of Unchahar was 27% and that of Talcher was 19% prior to take-over by NTPC. By undertaking a renovation and modernisation programme coupled with a sea-change in the approach towards operation and management of the stations, NTPC was able to dramatically improve the performance of these stations. For the year 1997-98, the PLF of Unchahar was 91% and that of Talcher was 52% reflecting the improved levels of performance. It may be noted that the additional generation from these units was possible with very low infusion of capital as compared to new capacity addition and thus, it is a cheaper and quicker route to augment the power supply position in the country.

The units of the Company over successive years achieved high levels of operational performance comparable to international standards of efficiency. NTPC achieved an overall Plant Load Factor of 75.20 during 1997-98 and an availability factor of 80.4%. The company's PLF could have been higher but for constraints of power evacuation in the Eastern Region.

In addition to the above, NTPC manages the Operation and Maintenance of Badarpur Thermal Power Station (705 MW) for the Government of India and the Balco Captive Power Stations (270 MW) for the Bharat Aluminium Corporation Ltd. The performance of Badarpur unit has shown dramatic improvement under the management of NTPC not only in terms of PLF, which increased from about 32% in 1977-78 to 72% in the year 1997-98, but also in terms of various other operation parameters.

Construction of Power Stations

NTPC, as a part of its capacity expansion programme to add 6270 MW during the IX plan, is presently undertaking the construction of 3170 MW capacity at an estimated cost of Rs. 10,159 crores. The Table 14.1 given below gives the details of NTPC's on-going projects.

Table 14.1 : NTPC's on-going projects

S. No.	*Project*	*Capacity (MW)*	*Location (State)*	*Primary Fuel*	*Approved Cost (Rs. /Crore)*	*Completion Schedule*
1.	Vindhyachal-Stage II	1000	Madhya Pradesh	Coal	2,754	Feb. 2001
2.	Unchahar-StageII	420	Uttar Pradesh	Coal	1,280	July 2000
3.	Kayamkulam	350	Kerala	Naphtha	1,311	Mar 2000
4.	Simhadri	1000	Andhra Pradesh	Coal	3,651	Dec. 2002
5.	Faridabad	400	Haryana	Gas	1,163	Jan. 2001
	TOTAL	**3170**			**10,159**	

*Table:14.1 *Out of this amount, Rs. 1,036 crore was spent in VIII Plan and the outlays in IX and X plan are envisaged to be Rs. 7,587 crore and Rs. 1,536 crore respectively.*

Consultancy Services

The corporation has a separate consultancy services division which provides consultancy services in all areas of power projects from concept to commissioning. The division provides consultancy in the areas of engineering of new projects, contracting and procurement services, operation and maintenance services, Renovation and Modernisation of Power Plants, research and Development Services and Training Services to SEBs/ IPPs.

FUNDING OF NTPC (COMPLETED AND ONGOING PROJECTS)

Sources of Finance

NTPC is a wholly owned Government of India Enterprise. Presently the whole of its equity is held by the Government of India. The company was formed with an authorised capital of Rs. 125 crore which as on 31st March, 1998 stands at Rs. 8,000 crore. As on March 31, 1998 the total investment of NTPC stood at Rs. 27,471 crore.

The company has been able to raise finances from wide spectrum of sources which includes multi-lateral/bi-lateral loans routed through the Government, external commercial borrowings, bonds issued in Indian capital markets, term from Indian Financial Institutions and Internal Resources. The company by dint of successful performance in Construction, Operation and financial Management, was able to inspire the confidence of international lending agencies like the World Bank, ADB, KfW, QECF, OPEC, SFD and various commercial banks/financial institutions who have extended financial support to the Company for its various projects. In addition, the Company was able to raise funds from domestic capital markets and also generate the requisite internal resources for financing its capital schemes. A measure of its success in operational performance and financial management is the fact that from the year 1989, the Company did not receive any net-budgetary support from the Government of India. The various sources of finance tapped by the company to fund its completed and on-going projects are discussed below:

External Sources of Finance

The total funds raised by the Company upto 31st March, 1998 from external sources are Rs. 14,266 crores out of which about Rs. 10,000 crores has been in the nature of external borrowing from multi-laterlal/bi-lateral sources routed through the Government of India and the balance Rs. 4,266 crores was raised by the Company directly:

Multi-lateral and Bi-lateral Sources

World Bank

Right from the inception of NTPC, the World Bank has supported several projects of the Company. The company's projects at Singrauli, Korba, Ramagundam, Farakka, Anta, Auraiya, Kawas, Talcher and NCTPP have been the beneficiaries of World Bank funding. The total utilisation of World Bank loans by NTPC upto 31st March 1998 has been Rs. 5,574 crores. Consequently, today NTPC is the single largest beneficiary of the World Bank funding (with an assistance of about US $ 4 billions) in the world. World Bank from time to time in their Appraisal/Audit reports has acclaimed NTPC's achievements in project construction within time and cost, initiatives in introduction of new technologies, operational performance and financial management.

Presently, the $ 400 million first Tranche of World Bank time Slice Loan is being utilised by the Company to finance the Vindhyachal Stage-II and Kayamkulam projects. This loan is directly made to the Company instead of being routed through the Government.

Asian Development Bank

The Second stage of Unchahar project taken over by NTPC from the UP State Government is party financed by the $ 160 million loan from ADB. This is also a direct loan to the Company from the ADB.

OECF Japan

The company's gas based project at Gandhar was partly funded through borrowing from the overseas Economic Co-operation Fund, Japan (OECF) for an amount of Rs. 1807 crore.

The OECF has committed till 31st March,1998 loans amounting to JPY 43 billion to part finance the on-going projects at Faridabad and Simhadri. The loans for Gandhar and Faridabad project were received through the Government of India, whereas the loan for Simhadri project is made directly to the Company.

Soviet Loans

The Company's projects as Vindhyachal Stage-I and Kahalgaon have been implemented with loans from the erstwhile U.S.S.R. amounting to Rs. 950 crores.

KfW, Germany

The loans from KfW, West Germany have been received for part financing the company's coal based stations at Singrauli, Korba, Ramagundam and Farakka, besides the gas station at Dadri. The total borrowing from KfW has been to the extent of DM 985 million.

Others

In adition to the above major sources of multi-lateral/bi-lateral assistance, the company received financial support from Saudi Funds, OPEC, Italian, U.K. and French Governments for its various capacity addition schemes. In respect of its erstwhile transmission line business, the company had also received Norwegian government funding. The total borrowing from these sources is about Rs. 650 crores.

External Commercial Borrowings

The successful track record of the company in construction and operation of power projects and its strong balance sheet have enabled the company to raise external commercial borrowings from a large number of foreign banks as syndicated loans or as export credits.

The first external commercial borrowing of the company was from consortium of banks lend by the Standard Chartered Merchant Bank for an amount of GBP 344 million to finance its Rihand Project.

The company had also raised export credit for its Kawas project from Banque Paribas, Paris for an amount of FF 1,539 million and from Banque Paribas, Belgium for an amount of BEF 1,250 million.

Further, the Company has also raised Buyer's Credit to part finance its Farakka project from a consortium of Banks led by the bankers' Trust Co., Hong Kong to the extent of DM 169.1 million. The outstanding balance DM 54 million from this loan was pre-paid by a fresh syndicated loan from consortium of banks led by Bank of America Singapore Ltd. to save on interest costs to the extent of Rs. 2.70 crores.

During the 1987-88, the company raised over JPY 100 billion through syndicated loans from the Industrial Bank of Japan and EXIM bank of Japan for filling up the gaps in funding requirement of various capital schemes. To take advantage of reduced interest rates prevailing in the Japanese markets, the Company subsequently raised JPY 23 billion from two consortiums of banks, led by Sumitomo Bank and JPY 15 billion from a consortium of banks led by Societe Generale Asia Ltd. to pre-pay the outstanding balances of the borrowings from Exim Bank and Industrial Bank of Japan. This has resulted in a savings of over Rs. 130 crores in interest costs.

Funds from Domestic Capital Markets

Power Bonds

NTPC was one of the first few companies which were initially permitted to tap the domestic capital market through issue of PSU bonds. The company made eleven Bond issues aggregating to an amount of Rs. 3,300 crores. This comprises of Taxable Bonds amounting to Rs. 1815 crores and Tax-free bonds amounting to Rs. 1485 crores. In terms of mode of offering, the Company made four public issues of Bonds raising Rs. 1,220 crores from the Indian public. While all the issues were over-subscribed, the last public offering of Bonds aggregating to Rs. 150 crores made by the Company in 1993, when bond market was sluggish, was over-subscribed by more than four times, which reflects the immense confidence enjoyed by the company among the investing public. The Company has also made private placement of bonds aggregating to Rs. 2,080 crores to various domestic banks and financial institutions etc.

Term Loans from Financial Institutions

NTPC raised term loans to the extent of Rs. 600 crores from domestic Financial Institutions such as Life Insurance Corporation of India, Unit Trust of India and HDFC. In view of large requirement of funds and also capability of the organisation to raise funds directly from capital markets as well as from external sources, in the past, this avenue has not been the preferred source of funding of NTPC projects.

Public Deposit Scheme

NTPC has also been accepting deposits from public to augment its resources for working capital requirement. The outstanding deposits as at the end of March, 1998 for Rs. 290 crores from about 18,000 depositors.

Internal Resources

The company has been able to generate internal resources required for its capital expansion schemes. Till 31st March, 1998, the company has utilised over Rs. 7,000 crores from out of its internal resources on its various capital schemes. After 1989-90 no budgetary support to the Company was available and it has funded all projects thereafter, either from generation of internal resources or from borrowings. Till 1993-94 Government of India did not insist for payment of dividend and permitted NTPC to deploy all its earnings for new capacity addition. However, under new guidelines of the Government of India, NTPC is now paying dividend since the financial year 1993-94. The last dividend provision made by the Company in 1997-98 is Rs. 503 crores.

Equity

The total equity of the company as on 31st March 1998 stood at Rs. 7546 crores which is fully owned by the Government of India. Out of this, an amount of Rs. 1,614 crore only was received as net budgetary support from the Government and the balance amount were received by Government of India from multi-lateral/bilateal institutions which were passed on to the Company as equity.

FUNDING PATTERN

The outlays of the company upto 31st March, 1998 on its various capital schemes has been Rs. 27,471 crores out of which Rs. 24,894 crores was spent on completed schemes, Rs. 2,341 crores is on its various on-going schemes and the balance Rs. 236 crores on Miscellaneous schemes. Figure 14.3 adjacent shows the funding pattern of the NTPC's completed, on-going projects and total investment.

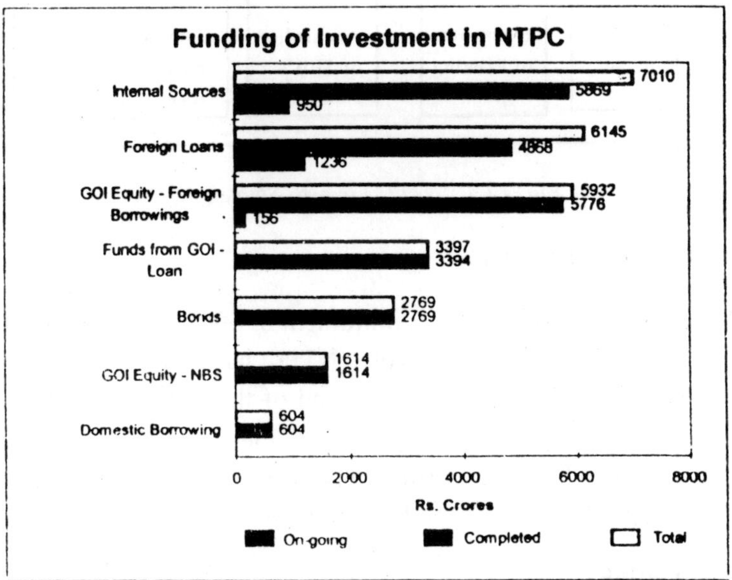

Fig. 14.3 : Funding of investment in NTPC as on 31st March, '98

GROWTH PLANS AND STRATEGIES

While maintaining its thrust on getting maximum out of the existing capacities, NTPC is also giving utmost priority to its future growth. NTPC is one of the nine PSUs identified by Government of India as 'Navaratna's capable of becoming global giants in their respective fields of operations. To guide its future growth strategies, NTPC set for itself a Corporate Vision to be one of the largest and best power utilities of the world. In tune with its Corporate Vision, the company has chalked out a growth strategy for capacity addition and selective diversification in related areas.

Multi-pronged Growth Approach

NTPC is adopting a Multi-pronged growth strategy i.e. capacity addition through Greenfield Projects, Expansion of existing stations, Acquisitions and take-overs, to accomplish its growth plans.

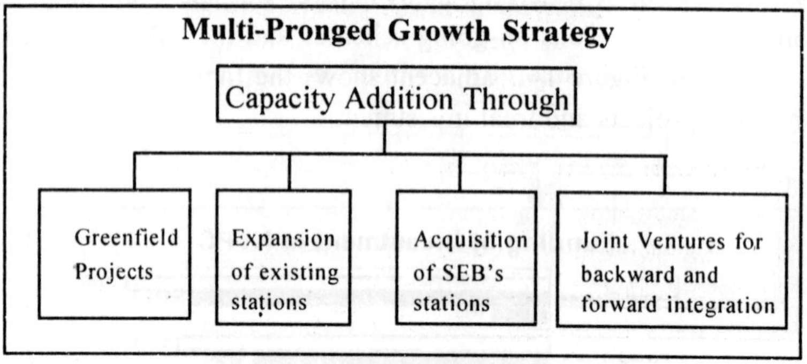

Fig. 14.4 : NTPC Growth Strategy

Greenfield and Expansion Projects

NTPC has fulfilled its capacity addition commitment of 5002 MW for the VIII Five Year Plan (1992-97). To maintain its growth rate, NTPC has drawn an ambitious plan of becoming a 30,000 MW plus company by the end of the X Plan period. In addition to 3170 MW under construction for completion in IX Plan, the company has identified a basket of new projects including both greenfield projects and expansion of existing stations for capacity addition of 3100 MW in the IX and 6,500 MW in the X Plan periods.

Selective Diversification

There are ample opportunities for NTPC to achieve required growth rate to become a global giant through new thermal capacity addition in India. However, associated with the best opportunity of thermal capacity addition in India, there are also certain potential environmental concerns/threat to capacity additions. These threats thus would need to be dealt with. Diversification will be one such key to the continued success. The menu of following new business areas would be comprehensively evaluated for diversification. Based on this evaluation, the most desirable business areas would be shortlisted/prioritised from the menu.

- Renovation and Modernisation of old power stations through a separate Joint Venture Company.
- Investments in LNG regasification terminal through a Joint Venture
- Investment in coal mining/coal washeries as backward integration for supply of coal to NTPC.
- Setting up of ash based industries to utilise fly ash in Joint Venture
- Setting up of small size pilot plants using renewable energy sources
- Setting up of hydro stations
- Acquisition of non-performing stations of SEBs against outstanding dues

FINANCIAL STRATEGIES FOR GROWTH PLANS

As per the recent Government policy Power Projects can have a debt equity ratio upto 80:20. NTPC, considering its low gearing, decided to take advantage of this Governmental policy and accordingly, the NTPC projects which have been approved in the last 2-3, years have debt equity ratio of 70:30, with the exception of Faridabad Gas Power Project.

The requirements of funds for capacity addition of 12,770 MW envisaged by the Company is 44,651 crores out of which Rs. 10,159 crores is in respect of on-going projects and Rs. 34,493 crores is in respect of new projects to be taken up by the Company. Out of Rs. 34, 493 crores required for new projects, the estimated outlay during the IX Plan is Rs. 11,270 crores, during the X Plan is Rs. 19,560 crores and the balance of Rs. 3,663 crores is required beyond the X Plan.

The requirement of internal resources for new projects is about Rs. 10,000 crores and the same is envisaged to be available to the Company despite difficulties in realisation of receivables from the SEBs. The balance requirement of funds being quite considerable, a mix of various suitable sources of funds is considered advisable as a strategy rather than pursuing with limited avenues of funds.

In formulating the financial strategy, the Company is guided by the fact that it has no foreign exchange revenues to provide a natural hedge for servicing external borrowings. Therefore, the Company would prefer to raise funds primarily from the domestic markets and would approach external sources only after exhausting possible domestic borrowing.

However, two exceptions are envisaged i.e. (a) Company needs to establish itself in foreign capital markets for which some foreign bond issues would be necessary; and (b) when the cost of borrowings from external sources is expected to be cheaper compared to domestic borrowings. The financial strategy is also based on the assumption that the loans now being taken by NTPC will not have the Government of India guarantee and will have recourse only to NTPC balance sheet. Accordingly, the following financial strategy appears to be most suitable and feasible.

Domestic Sources

(i) Domestic Bonds

The bonds issued by NTPC in domestic market have always been received very well by market. Keeping in view our past experience of public issues/private placements, it can be safely assúmed that bonds to the extent of Rs. 1000 crores every year (maturity ranging 7 to 10 years) can be easily issued by NTPC in domestic market to finance the new projects.

(ii) Funding from Banks and Indian Financial Institutions:

The Company finds that the response of domestic banks and financial institutions is very encouraging to fund its capacity expansion programme. The Company would like to raise funds from this source, as well, in case the overall cost is comparable to cost under a Bond issue.

External Sources

(i) Direct Market Borrowings

The external capital markets in United States of America, Europe and Asia have largest availability of resources and good corporate can expect to raise substantial funds from this source. It has been considered desirable that NTPC should have capability to approach the foreign capital markets directly for raising debts. This should progressively reduce the share of multilateral funding and increase direct borrowing from market which could be raised regularly and from time to time depending on progress and need of funds for the projects. As a first step in this direction, NTPC has gone for credit rating from Standard and Poor's and Moody's. It is proposed that NTPC would make a maiden bond issue in external markets during 1998-99 in order to establish its presence in the foreign capital markets followed by similar issue in 1999-2000. It should enable NTPC to become a known name in the international capital markets and would enable it to tap this segment depending upon the requirement of funds and opportune market conditions.

(ii) Syndicated Borrowings

NTPC had not approached the foreign loan syndication possibilities for some time. During the year 1996-97 we made a borrowing of JPY 5 billion from international market for re-financing an earlier loan which wás on higher rate of interest. Similar external borrowings to the extent of about JPY 33 billion and DM 54 million have been finalised in 1997-98 for refinancing pre-payment of earlier loans. It is now proposed that NTPC should regularly approach foreign banks for syndication and this should form a regular part of our financing plan for new project.

(iii) Co-financing along with World Bank

In the area of syndication of loans through foreign banks, it is also proposed to explore the possibility of co-financing along with World Bank. However, adoption of this route shall depend an alternative cost of borrowings. This matter needs to be discussed further with World Bank.

(iv) Multilateral Fundings

NTPC will continue to pursue with for funding by Multilateral Agencies like World Bank, ADB and OECF as part of its efforts to part finance the projects. However, it is envisaged that compared to the earlier projccts, the ratio of borrowing from this source will get reduced substantially with increase in borrowing from foreign capital markets and banks.

The possible scenario of funding of the new projects based on above strategy is expected to look as given in Figure 14.5 below:

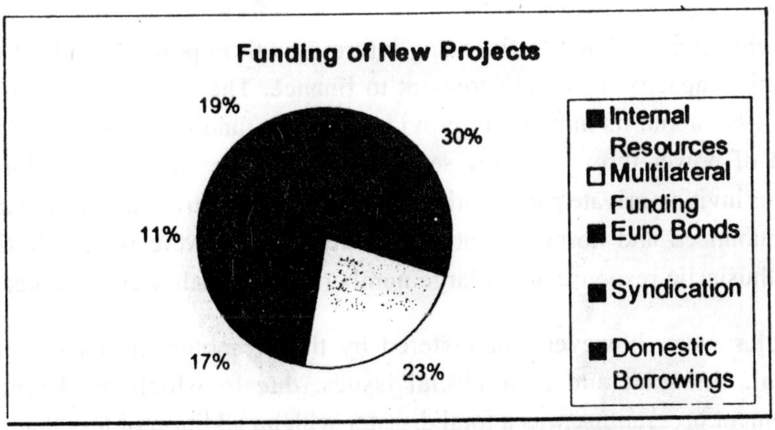

Fig. 5 : Funding of New Projects of NTPC

NTPC is confident that with the above proposed financial strategy it would be able to take up and execute the new projects under consideration with a suitable mix of different types of sources of funds in order to become a 30,000 MW company by the year 2007.

FINANCING OF INDEPENDENT POWER PROJECTS : A REVIEW OF MAJOR CONSTRAINTS

15

A.K. Sah
President, Premier Energy Agency, New Delhi

The Fourth National Power Plan, 1997-2012 of the Government of India has projected capacity addition of more than 150,000 MW in the next 15 years upto the end of Eleventh Five Year Plan. In addition, bulk inter-regional transmission lines have to be provided to facilitate flow of power across regions to achieve optimal utilisation of generation resources. On an average, therefore, around 50,000 MW capacity has to be added in each five year plan period when almost the entire power development was the responsibility of the public sector.

Taking a rough estimate of Rs. 4 crores per MW for generation projects and Rs. 2 crore for transmission and distribution, the investment requirement in each Five Year Plan period will be of the order or Rs. 30,000 crore against an expenditure of less than 80,000 crore during the Eighth Plan period.

The challenges related to financing future growth in power is indeed formidable and beyond the capacity of the government to finance. The Government had realised as far back as 1991 about its inability to provide adequate funding for meeting the fast growing demand of power through public sector sources alone and had therefore announced a policy for inviting private participation. A series of legislative and administrative measures were announced and norms for attracting investments were issued. Initially there was very enthusiastic response and a large number of proposals were received.

Hurdles were, however, encountered by the promoters in sorting out several key technical, financial and commercial issues, due to which the Eighth Plan power development programme was a total disaster with an addition of less than 1700 MW new generating capacity, and almost zero addition in the private sector.

One of the main deterrents which has held up private investment in the power sector is with regard to the poor financial health of the State Electricity Board. Several studies carried out by international consultants have recommended creation of separate generation, transmission and distribution companies and tariff regulatory commissions in each State, so as to delink the electricity sector from political influence and make them commercially viable entities.

The question which is uppermost in my mind is whether restructuring alone will lead to the desired financial health of SEBs? Orissa has gone ahead and done it—With what result? The initial experiment with management of distribution system by a private sector utility was not successful and had to be terminated. A recent newspaper report states that Noida power company has also not been able to provide satisfactory power supply to consumeRs. The causes for these failures can be quite different but the clear message is that there has not been an attitudinal change either in the functionaries of the government or the private sector promoters themselves or the consumers who have for the last fifty years been brought up on the concept that it is the responsibility of the government to make available electric supply to them at no cost. The opposition by a number of States to the Power Regulatory Ordinance is issued by the Government of India is an eloquent testimony to these political compulsions so much so that a modified much watered down version has now been legislated. A Planning Commission estimate is that effective power subsidy to the agriculture sector would have been Rs. 17,491 crore in 1997-98. This represents an increase of over Rs. 2100 crore in one year. The result is poor financial health of State electricity boards,who are neither able to maintain their assets properly nor invest in new ones to improve efficiency. The tariff regulatory commissions which will now be formed under the new act will have no teeth to enforce its decisions.

There is no doubt that rural electrification and supply of power to the agricultural sector for food production is an essential requirement for the country's development. It is also clear that rural electrification and agricultural supplies cannot become commercially remunerative when even light and fan loads and domestic power supplies in urban areas are unremunerative.

I have since long advocated that rural electrification work should be delinked from the commercial activities of State electricity boards and taken up as a part of integrated development activity of the rural areas by the State government. For this purpose institutions like NABARD and REC may join hands together and provide funds directly at concessional rates to Rural Electrification Cooperative societies or any other such bodies, which may be set up in each district. The electricity boards may be responsible for supplying bulk energy to these Cooperative Societies at reasonably remunerative tariff to be determined by the Tariff Regulatory Commission. By removing the major loss making operations from the purview of SEBs, the wholesale restructuring of the electricity sector at the State level may not be immediately necessary.

The fact that NTPC which is a public sector undertaking has been ranked third Internationally in terms of generation efficiency explodes the myth that public sector undertakings cannot function efficiently and only privatisation will bring the desired improvement in performance. My own long experience in the Indian power sector and studies of various systems abroad leads me to conclude that it is the competitive environment which includes any undertaking whether in the public or private sector to improve its performance. In a monopoly situation private sector organisation could behave worse than a public sector undertaking. I do not mean to say that private investment in power sector should be discouraged. This has to be brought in to achieve the desired level of economic growth but a healthy p.s.u. will act as a restraining force on unbridled working of private sector utilities purely on profit motive and will go a long way in building investor confidence for attracting investments in this sector.

Apart from this issue of wholesale restructuring of the electricity sector at the State level, there are other risk factors which act as deterrent for private investors:

(a) Undue delays in project clearances and achieving financial closing:

Although the States have been given the powers to clear projects upto 1000 MW capacity on competitive bidding basis, most of the States have not developed the required infrastructure and expertise for evaluation and clearance of the projects. This has brought traders and power brokers in the field who corner the projects without doing adequate technical and financial homework and thereafter try to find buyers at a premium.

The system for bid invitation and their evaluation has not been systematised. The SEBs expect the bidders to do complete technical and commercial evaluation of the project, tie up their partners and submit bid binding themselves to project cost and tariffs. This is clearly impracticable and has led to non-serious speculative bidding. It is necessary that full project details be given in the first instance to bring in transparency in evaluation and award of project. Until this is done, avoidable delays will continue to take place with corresponding escalation in project cost.

(b) Escrow Account and State Government Guarantees:

Since most of SEBs are commercially unviable organisations, outstanding dues of more than Rs. 10,000 crores have accumulated as outstanding dues to Central Sector undertakings. The estimated cost of power purchased by the SEBs and other electricity undertakings during the year 1996-97 was of the order of Rs. 17,000 crores and the commercial losses sustained by them was Rs. 10,000 crores. Therefore, any Escrow account and Letters of Credit issued by SEBs will be at the cost of some other essential revenue expenditure. It is for this reason that various financial institutions are examining the revenue accounts of SEBs and have questioned their capacity to open the Escrow accounts for the number of projects for which they have finalised the offers. There being no system of merit order rating of these projects, those who are more active and command political influence are able to hold back finalisation of project clearances till their terms and conditions are met.

(c) Fuel Supply the Transportation Agreements

Assured fuel supply for any power generation project is a must. Fuel supply agreements have become the single biggest bottleneck in achieving quick financial closer. No investor is prepared to take fuel supply risk. The generation tariff norms which have now been fixed on availability of the power plants at a high level of 80 to 85% presupposes that right quality and quantity of fuel is regularly made available to the power plant. As far as my knowledge goes, not one legally binding fuel supply agreement with full responsibility for its delivery at the power station siding has so far been executed. Both Vizag and Bhadravati projects have been inordinately delayed on this account. I can cite many other projects which have reached an advanced stage but in which clear agreement on coal quality is not yet available. This is a clear case in which due to the conflicting inter-departmental interests and various restrictive laws etc., the Government has not been to enunciate a clear policy for fuel supply.

(d) Political and Regulatory Risks

The political risk perception for India is high. Although successive governments have reiterated the policy of liberlisation, in actual practice clearance of projects through the State government departments and various Ministries of the Government of India is very time consuming as well as frustrating. Even the fast track projects which should have been nearing completion have not yet received all the clearances and guarantees. Legal issues relating to validity and enforceability of guarantees take a long time to be sorted out because of multiplicity of governmental authorities with dispersed responsibilities and lack of clear legal framework and absence of quick dispute resolution mechanism.

The potential effects of variation in inflation rates and fuel and operating costs along with foreign exchange availability and convertibility and poor credit worthiness of the purchasers of power are other major irritants due to which several projects have met with premature deaths.

The question of foreign exchange risk insurance is one of the major stumbling blocks in many international IPP negotiations. Different governments have developed or are developing country specific policies, the interpretation of which have to be precisely understood. Changes in duties levied on imported plants or in taxation on the contractor's business in the host country or changes in local employment conditions can have major cost implications.

(e) Contract Time and Cost Overruns

The gestation period of power projects being long the risk of contract time and cost overruns due to causes beyond the control of promoters is quite high. Establishing that such time and cost overruns can legitimately be added on to the approved project cost is a very difficult and tricky issue which often leads to long drawn arbitration and legal proceedings. If, with the setting up or regulatory commissions, this procedure can be simplified and made less time consuming, it will help in building up investor confidence. The Indian Government has announced several steps for initiating reforms in the existing power scenario. It this endeavour the Government of India is focusing on the following.

— Providing funds to public sector projects

— Expediting reforms related to transmission and distribution and tariffs

— Encouraging private sector participation

— Promoting mega projects

— Speedy clearances of pending projects through a system of rigorous monitoring with assigned responsibility.

— Providing counter guarantees to upcoming projects.

The new reforms also include the Central Government's proposal to delegate power to the State Governments to decide FDI cases involving an amount up to 1,500 crore rupees for funding 1,000-1,100 MW projects.

These are commendable steps. Time alone will show to what extent these measures really translate into megawatts during the Ninth Five Year Plan and relieve the power consumers in general from the day-to-day harassment of power cuts, and poor quality of power supply.

STATUS PAPER ON THE ACTIVITIES OF BHEL WITH SPECIFIC REFERENCE TO THE FINANCING NEEDS

16

B. Bhambani
General Manager, Project (Finance)
Bharat Heavy Electricals Limited

ACTIVITIES OF BHEL

Bharat Heavy Electricals Ltd. (BHEL) is a Public Limited Company with 68% equity held by Government of India and 32% equity held by financing institutions mutual funds, insurance companies, foreign institutionals investors and employees.

The first plant of BHEL was established nearly 39 years ago at Bhopal and was the genesis of the Heavy Electrical equipment industry in India. BHEL is today the largest Engineering enterprise of its kind in India with well recognised track record of performance, making profits continuously since 1971-72. It achieved a sales turnover of Rs. 64500 Million with a pretax profit of Rs. 8644 Million in 1997-98.

BHEL caters to core sectors of the Indian Economy viz, Power, Industry Transportation, Transmission, Defence etc. the wide network of BHEL's 14 manufacturing divisions, nine service centres, four power sector regional centres, and over 150 project sites enables the company to be closer to its customers and provide them with suitable products, systems and services at competitive prices. Having attained ISO 9000 certification, BHEL is now embarking upon the total quality Management.

The Company's inherent potential coupled with it strong performance over the years, has resulted in its being chosen as one of the "Navaratna" PSEs which are to be supported by the Government in their endeavour to become future global playeRs.

BHEL's operations are organised around three Business sectors viz, Power, Industry and International. Power Sector comprises Thermal, Nuclear, Gas and Hydro Power Plant business. Today BHEL supplied sets account for nearly 54681 MW or 65% of the total installed capacity of 84810 MW in the country as against NIL till 1969-70. BHEL has proven turnkey capabilities for executing Power Projects from concept to commissioning. It possesses the technology and capability to produce thermal power plant equipment upto 1000 MW rating and Gas Turbine Generator sets upto a unit rating of 150 MW. Cogeneration and combined cycle plants have also been introduced to achieve higher plant efficiencies to make efficient use of the high ash content coal available in India. BHEL supplies circulating fluidised boilers to both thermal and combined cycle power plants. BHEL also manufactures 235 MW and 500 MW Nuclear sets.

Custom - made huge hydro sets of Francies, Pelton and Kaplan types for different head-discharge combination are also engineering and manufactured by BHEL. In all more than 720 sets of Thermal, Hydro, Gas and Nuclear have been given on contract to BHEL as on date. The power plant equipment manufactured by BHEL is based on contemporary technology comparable to the best in the world and is also Internationally competitive. BHEL have proven expertise in renovation and maintenance of variety of Power Plant equipment besides specialised know-how of residual life assessment, health diagnostics and life extension of plants. To take advantage of technology advancement in this area BHEL has formed two joint venture companies, one with Siemens Germany and the other with GE-USA to take up on turnkey basis contracts for renovation and modernisation of Coal based stations and Gas Turbine stations.

In the field of Industry sector, BHEL is a major contributor of equipment and systems to Industries like cement, fertilizers, refineries, petrochemicals, steel, paper and telecommunication. The range of systems and equipments supplied include captive power plants, high speed industrial drive turbines, industrial boilers and auxiliaries, waste heat recovery boilers, Gas Turbines, heat exchangers and pressure vessels, electrical machines, pumps, valves, seamless steel tubes, process controls etc. BHEL is a major producer of large size thyristor devices. It also supplies digital distributed control systems for process industries and control and instrumentation for power plants and industrial application. BHEL has the capability to make Simulators for power plants, defence and other applications.

In addition BHEL supplies a wide range of transmission products and system upto 400 KV class. These include high voltage power and distribution transformers instruments transformers, dry type transformers, SF6 Switchgear capacitors, etc. For economic transmission of bulk power over long distance, High Voltage Direct Current (HVDC) systems are supplied.

A high percentage of the trains by Indian Railways are equipped with BHEL's traction and traction control equipment including the metro at Calcutta. BHEL supplies 3900 HP broad gauge AC locomotives to the Indian Railways and Diesel shunting locomotives to various Industries. BHEL have leased 53 nos 5000/4600 H.P. AC/DC locomotives to Indian Railways during the years 1996-97 and 1997-98.

BHEL has effectively participated in the National programme for harnessing Non-conventional Energy Sources. BHEL supplies wind generators, solar power based pumps, lighting and heating systems are operating at various locations. BHEL has also commenced manufacture of large Desalination plants to help arrangement the supply of drinking water to people.

In the field of exports, BHEL products, services and projects have been exported to 52 countries ranging from USA in the west to New Zealand in the far east. The cumulative capacity of power generating equipment supplied by BHEL outside India is over 300 MW.

BHEL has well established R&D centre at Hyderabad and this in turn is supported by product based R&D centre at each of the manufacturing divisions. Several State-of-the art technologies developed in-house have been commercialised like FBC boilers, Direct ignition of Pulverised Coal (DIPC) etc. BHEL is upgrading its products and related technologies to contemporary levels through continued in-house efforts and acquisition of technology from leading Engineering Organisations of the world.

FINANCING NEEDS

BHEL is able to fund from its own resources the Capital Investment of about Rs. 100-150 crores per annum. If and when necessary BHEL can raise funds on its own financial strength and good credit rating.

BHEL's present requirement of support from Bank/FIs is only for projects of its customers, both public as well as private independent power projects and captive power projects of the industry. For BHEL to obtain orders for Boilers, Turbo Generators, Gas Turbines from IPPs and industry is based on its capability to syndicate funds from IFIs/ Banks/FIs and Foreign banks at attractive terms for borrower.

In line with extent policy of Government of India for financing of IPPs and as modified by All India Financial Institution, the permitted pattern for financing of an IPP is based on Debt : Equity ratio of 70:30. Indian FIs including PFC and commercial banks generally provide rupee resources to the extent of 40% of the project cost. These funds are normally used by IPPs for :

i) Rupee outlay on taxes and duties.

ii) Insurance

iii) Civil Works

iv) Erection and Commissioning

v) Minor indigenous equipment which are essentially procured within the country.

The above scope generally constitutes about 30-40% of the project cost Promoters of IPPs demand that supplies of main equipment i.e. Boiler and TGs which is nearly 50% of EPC bid should arrange/provide credit to the extent of 30% of the project cost. If BHEL has to take up EPC package of an IPP, it has to syndicate financial package from IFIs/FIs to cover total cost. While IFIs can provide rupee resources to the extent of 40% of the project cost, the balance 30% has necessarily to come from FIs. Presently commercial credits from Foreign banks are not available easily for maturity periods beyond six to seven years. In comparison to local EPC contractor, foreign bidders have the advantage of arranging funds from FIs on soft export credit terms as their supplies are considered as export from the country offering the credit. The present GOI guidelines governing the flow of rupee resources from IFIs for an IPP restrict BHEL's participation in EPC bids for private power projects.

It would not be possible for BHEL to provide such large amounts as suppliers credit from its own balance sheet become internal resources generated by BHEL are just about adequate to meet its own fund requirement. World over equipment suppliers do not provide suppliers credit from their own balance sheet. In view of easy availability of export credit, most of the IPPs have preferred to procure their main equipment from overseas companies, backed by such soft credits.

BHEL and indigenous Industry is suffering a serious disadvantage in exploiting emerging possibilities for supply of man equipment to IPPs despite being competitive as the present Government of India's guidelines restrict rupee resources from IFIs upto 40% of the project cost.

Which BHEL has taken various measures to strengthen its capability to arrange project finance, it is necessary that Government of India's amend its guidelines and permit IFIs to provide rupee resources to IPPs upto 70% of the project cost on the condition that project will use 100% endogenous equipment and developer will not have clause in the PPA for exchange risk as pass through to the utility. Such a arrangement will assist utility to negotiate a very competitive rate for purchase of power from IPPs. Alternatively EXIM-BANK should be authorised to provide buyers credit to IPPs at soft rates in addition to funds being provided by IFIs as supplies to IPPs are considered as deemed export. The credit from EXIM-BANK to IPP can be supported by ECGC for covering risks during construction period.

ACCESSING DOMESTIC CAPITAL MARKETS

17

T.N. Thakur
Director (Finance), PFC

The funding base of PFC is 14% of equity, but external loans of 17% , the government of India loans 17%, mostly the Rupee loan which government of India gives us out of the ADB and the World Bank loan proceed and domestic 33% comprising largely of tax free or SLRC bonds; and some taxable bonds. But, years to come PFC will have to depend more and more on commercial funds that are soft funds. For resource mobilisation, whether from the external markets or the domestic markets, PFC has obtained credit ratings for Moody's. PFC has got fixed deposit scheme rated by the domestic rating agency who have given it 'AAA' rating. PFC has not got bonds rated because these bonds are only on the private basis for which rating is not compulsory. PFC has also planned to undertake public issue in the near future. PFC offers term loan both in terms of foreign exchange as well as rupee. When PFC gives loan, it helps the borrowers to manage the exchange rate effectively. PFC has given both the options to borrowers; (i) they can take loan from PFC in rupees, even out of the foreign currency borrowings and on the rupee loan, it will be normal lending rate; (ii) they can also take foreign exchange loan from PFC. Then, it will be back to back with the margin of PFC. How is that managed? PFC has set up an exchanged risk administration fund to which six monthly contribution to be made by the borrowers, if the borrower has taken the risk and if PFC taken contribution by PF. Our experience has been that these six monthly contribution are required to be made by the borrowers or the PFC. According to the set formula his contributions are worked out to be less than the market cost. Therefore, the cost to the borrower or cost to the PFC is lower than if they had gone to the market. And yet the fund has been growing. The fund has assumed a size of Rs. 300 crore. This Rs. 300 crore is invested in the market so that the fund carns for itself. It is managed by a group of trustees and that PFC cannot be a beneficiary for this fund. For example out of this Rs. 300 crore PFC can invest in ICICI Bonds, IDBI bonds, but not in PFC Bonds. That has been done to bring in an objectivity in the investment of funds so that the funds get maximum aid which is possible in the market. Now the contribution we have been working out, initially the contribution worked out to be about 78%, but today the contribution is less than 4%. And I do not think that anywhere in the market, you can get handling cost of less than 4%. PFC feels that over a period of time the fund itself will be self-sustaining; and will be able to make good at the losses without further contribution to the fund.

But, there is a problem in the sense that this fund can take care of the excess losses if the depreciation is not more than 6, 7, 8% a year on an average. But, if the depreciation goes beyond 10% a year average, then, may be the contribution will be required to be made more. But in that case the handling cost will also go up; I mean that foreign exchange risk that one has to bear. PFC also provides a part from term loans. It provides payment guarantees for fields. PFC provide leasing, loan syndication for the private power projects and for the SEBs projects and also gives the working capital loans for short duration.

PFC is considering to provide advisory services to clients if they so need. The first operational parameters for PFC any SEB or State generating company, to become eligible for PFC funding is to sign what is called as Operational Financial Action Plan (OFAP). If they do not have a OFAP, PFC does not lend. But, this OFAP has milestones with regard to different financial operations parameters. OFAP is monitored by PFC. It is true that quite a number of SEBs canot be lent money immediately because of their poor financial health. But through the OFAP mechanism PFC tried to make projections for their future growth and then lent money and tried to secure PFC money in that manner. And, wherever the default on OFAP, that is considered to be default on the loan. PFC has a very tight project appraisal and financial appraisal. I can tell that once some project or some NTPC project has been able to pass the test of PFC, I am sure that scheme will not land in trouble; because PFC looks into every aspect of it. PFC has expertise because it has been looking at the power sector for the last 10 years. The terms and conditions of the loans granted are not tough, but quite pragmatic.

PFC also takes a real life view of anything. It is not that PFC is totally inflexible. It is flexible. PFC does credit rating of each electricity board for internal purposes. On the basis of the credit rating, PFC determines the exposure limit. While PFC does not distinguish between two boards as regards the rate of lending, but it does distinction between two electricity boards as regards the extent to which PFC can take exposure. If there is a creditworthy board, PFC takes larger exposure with them. If the board is not so creditworthy, PFC take small exposure with them. That distinction is made. There is no distinction as far as the rate of interest goes. PFC does not make a distinction between the private sector project, which is good and State owned project. However, PFC has an obligation because of large equity participation by the government and therefore, PFC gives preference to the SEBs and the State sector projects.

In terms of prioritisation of the project, we give more priority to transmission and distribution. PFC's prioritisation is on the basis of lower lending rates for the project that it considers are of high priority. Experience has shown that inspite of huge investment in the generation sector, there are areas where power is not reaching at the consumers' doorstep because of bad system, because of problem of transmission and distribution. It is projected that the eastern sector is surplus in power. PFC has a tight security arrangement and that is the reason there is no default inspite of the sick electricity boards, 98% collections are on due date. This is the projection PFC has made in regard to disbursement. This is based on PFC's vision of having a CRF of atleast 5% in the private sector funding and CRF of 20% in public sector funding. On that basis PFC has made a projection according to which by the year 2012 we should be able to disburse Rs. 5,500 crores a year. We are not going very fast. In 1994-95 PFC disbursed Rs. 800 crore, 1997-98 PFC disbursed Rs. 2,100 crores. That is the growth pattern of PFC in regard to market borrowings. PFC is concerned about the asset liability match and, therefore, PFC realises that it has to borrow larger maturity fund. PFC gives moratorium up to six months after the date of commissioning in all our loans, PFC is giving moratorium up to 6 months after the date of commissioning and repayment period varies from 8 to 10 years. In the market you do not get funds for 13-14 years, maximum is 7 years. That is why PFC tried to go to offshore markets where you get longer tenure loan. First time PFC went for a 5-year loan. When PFC went for 12 years' EURONET and it thought it would be an issue of 20 years maturity. But, by then the South Asian crisis began and the market was very bad for Indian issues. Therefore, PFC could not go to the market at that time. We have again started. You would have noticed PFC issues have foreign currency loan and from the capital market was the only major issue from India after the South Asian crisis. When PFC was in the syndication stage of loan, Pokhran tests took place, sanctions came and Moody's downgrading game came. Yet PFC raised the money at the price announced and on schedule.

That shows that people in the capital markets, whether it is offshore or Indian, are quite keen to make investments in good credits. PFC has to do its homework well. It has to convince the lenders that there is hope, possibility of making good use of their investments.

ACCESSING FINANCIAL MARKETS : ONGC'S EXPERIENCE

18

I.N.Chatterjee
Director (Finance) ONGC

The International capital markets are generally dominated by various debts and equity instruments. In the past ONGC had accessed a large number of markets, both in US and Europe; and various other bond markets, covering the Japan Bond, the Swiss Bond and Eurobond. Basically, when ONGC considered accessing resources from the international market, it had to look at both the debt, equity security. Now in the U.S markets, there are two basic routes for non-government, non-US corporate fund raising. All security exchange covers registered route or Rule 144A route under which there are specified institutions known as qualified institutional buyers. These can be accessed and ONGC need not go through ACC registration and generally these QIBs, which own atleast US $ 100 million of special type of security. Otherwise one cannot access that market without going through the ACC routes. In the case of institutions in Japanese markets, or German markets the compliance requirements are not so rigid as in US market and procedure is simpler. ONGC prefers to go to the Japanese market, if appropriately rated through one of the institutions, particularly the Japanese bond institution. That rating helps. It is observed that during the period 1990-97 there were about 35 institutions of fixed and floating rate international bonds from the Indian issuer and overall US $ 4.7 billion amount from various international bond markets, out of which ONGC has accessed about 10.82% of the total amount raised during this period. Whereas the syndicated loan market is much wider and larger, during the corresponding period 370 loans have been transacted, which has raised approximately US $ 11.5 billion.

ONGC did not have much need for accessing the resources from the domestic market, barring some term dealing in 1993-94, for which a couple of commercial banks were approached. ONGC also accessed FCNRB route because during that time FCNRB was a cheaper option, particularly short term fund as compared to routing resources from various banks. So even today possibly that could be one of the options available, for raising resources from FCNRB market, because lot of banks have resources. But with the foreign premium presently available, one has to look into the cost-benefit aspect and see as to which would be the cheapest.

With the current downgrading in the rating, one has to see as to which way the domestic market resources can be accessed.

A lot of facilities are available with the foreign banks. Since ONGC is in a comfortable position, it did not go through this route, except or the short duration during which period it accessed about Rs. 2,000 crores; and at that time ONGC accessed FCNRB markets and raised about US $ 5 million .

IDBI FUNDING OF POWR PROJECT

19

A.K.T.Chari
Coordinator, IDBI

IDBI is a leading financial institution in the country. As a consortium with other institutions like ICICI, IFCI, UTI, LIC, SBI and other commercial banks, it financed a number of power projects.

Since 1993-94 provided financial assistance has been provided to 32 IPPs with an aggregate capacity of about 13,000 MW. Of these 5 IPPs with a capacity of 1,274 MW have already been commissioned. One of 650 MW is partially commissioned; 10 with the installed capacity of 3,600 MW are under implementation. Out of these it is expected that atleast 2,000 MW will come into operation during the current year. 14 IPPs with installed capacity of 6,481 MW are under various stages of negotiations and partial financial closure has been reached. Two projects are still under re-negotiation with the Government of India in Ministry of Finance and Ministry of Power. Last year IDBI negotiated about 11 power projects with 4,777 MW capacity and agregate project cost of Rs. 18,600 crores. Total cost of all the 32 projects is around Rs. 54,000 crores; and the institutional assistance was about Rs. 19,000 crores.

The large portion of assistance has gone to four States, Andhra Pradesh, Madhya Pradesh, Gujarat and Tamilnadu. A number of projects are under negotiation. Even though the financial institutions have sanctioned financial assistance and there are various other procedures which are being negotiated now. Of the 32 projects already sanctioned, it can be stated that fuel wise categorisation is as follows: coal and lignite occupy the largest proportion of about 60% and then the liquid fuel based around 5500 MW, which include naptha gas, LHS oil. As a base load station IDBI has certain reservations on liquid fuel projects because they are subject to forex fluctuation.

Our main concern is financing the power project is the financial health of SEBs. On many forums, it has been talked about reforms are being set in motion. On every forum experts talk about SEBs are becoming weeping boys, but it is not a right situation. Many of the SEBs operate plants, almost at 95% PLF. The availability in some of the plants is 100%, which means that this far exceeds to some of the IPPs, which are giving guarantees in the new scenario. It is not correct to blame entirely the electricity boards for the yields, but the reason is somewhere else. The problem with the Electricity Boards is the agriculture subsidy. There is a tendency for larger creamy customers to going for the captive power generation which will aggravate further the financial position of SEBs. What will happen when the creamy consumers set up the captive power generating plants. And to that extent your percentage of the agriculture consumption goes up on the overall consumption.

Reforms have been discussed for quite some time. It is making more headway now and CERC Bill has been passed by the Parliament and subsequent to that we feel that State Electricity Regulatory Commission also will be set up. Looking into the financial position of the SEBs, there is only a short term solution now and that is the ESCROW mechanism, that ONGC insists on all the SEBs. But, the problem is that the total ESCROW capacity, is quite limited. The ESCROW capacity is based on their current revenue generation. Future increase in the tariff is to be viewed. There is going to be any increase in the agriculture is not certain now. Assuming that the State Regulatory Commissions are able to do their job properly (and the State expects the mandate) what might happen. The tariff may go up at that point of time, probably the ESCROW mechanism can be re-visited and one can take a fresh look. But, currently there is no option except to go for ESCROW mechanism in view of the financial health of the SEBs.

IDBI faces another problem. A number of project proposal receive signed MOUs or bidding process. Capacity is seen in access of what they can really afford in the sense for which stage can offer ESCROW. Since ESCROW capacity is limited, some of the projects may be falling in the way side, so that needs to be addressed.

The Planning Commission has projected that all the 19 SEBs incurred subsidy of approximately Rs. 18,000 crores. during 1996-97. In the newspaper report on July 15, 1998, it has been mentioned that the total subsidy to agriculture and domestic consumers will be as high as Rs. 25,000 crores during the current year (1998-99). That includes not only power but includes other sectors, the fertilizer and the other issues. Now the quantum of other subsidies. In 1996-97 the subsidy to agriculture was almost Rs. 14,000 crores; to domestic sector it was around Rs. 4,000 crores, making a total of Rs. 18,000 crores There was a receipt from the government upto Rs. 2,400 crore and the cross subsidisation that the HDC consumer pays. So, still uncovered subsidies are of Rs. 8,200 crores.

The other issue is of fuel and transportation. The liquid fuels are generally imported because the country is still not self-sufficient and SEBs cannot absorb the high cost of power and in case of violent forex fluctuation, there is one of the risks which you have to face. The fuel cost has to pass through most of the power purchase agreement. This might create some problem in the future. What I mean is the high cost of power, the SEBs may ask the liquid fuel based projects to bagged down to the minimum of 61.5 which is in the rights of the PPA obligation. The Railways do not have necessary infrastructure. Neptha is a class 'A' material and is highly explosive and the railways do not take transportation risk. The first such acceptance of risk by the railway will be probably from the one power project in Andhra Pradesh and that might become a model fuel supply agreement for coal.

We have a problem with the evacuation system of the IPPs. Large power projects, of 500 MW and above, need large expenditure on the T&D system. The State Electricity Boards sometimes are not adequately providing for such a T&D system. And one of the issues under the PPA if the T&D system is not ready, the SEBs will have to pay on the deemed generation.

Normally it is expected that 60% of the entire power sector investment goes in generation. But in the Eighth Plan T&D was allowed only 28% as against 72% in the generation. Normally T&D required 40%. That means that there is some kind of negligence on the fund flow into the T&D sector.

T&D losses are very high and probably the figures of 20-22% cannot to say with certainty. A lot of pilferage is taking place. Each 1% reduction in the T&D Loss will save approximately 700 MW of additional capacity requirement.

While working out the issue of ESCROW capacity IDBI normally asks and assurance of 1.25 times the loan requirement. Suppose if the monthly bill of the IPP to the electricity board is around Rs. 20 crores, then it is suggested that Rs. 25 crores monthly ESCROW should be available so that in case there is any shortfall in the revenue fund collection or from some customers, that will be taken care of. We also suggest that ESCROW should be provided to 80 to 85%, whatever be the provision in the purchase agreement. On the new power projects, they cost Rs. 3-4 crores MW and it will be essential to operate them at 68.5% PLF. Once has to seriously think operating a plant at 68.5% of such a high cost plant. Some of the Electricity Plants are already operating, Andhra Pradesh operating on 77% PLF. It is a point for discussion.

The third, IDBI suggests that the ESCROW funds should come from the dedicated circle so that the fund flow can be properly monitored and there is no litigation between the two IPPs or SEBs and IPPs. This will ensure for that the funds are flowing through the ESCROW mechanism and all the lenders as well as the companies are getting their money. The ESCROW itself is a highly structured mechanism. The ESCROW funds immediately goes through the trust account from where the money flows through on a priority basis. First it goes to the operation of the plant itself; then it goes for wider expenses the fuel expenses; then comes the interest and principle; and finally the money flows out to the company. The ESCROW analysis is a very involved job. It depends on how fast the SEBs are implementing the reforms? It can be done over a period of five years or six years. So you cannot go on financing the projects indefinitely. Theoretically you can finance any type of electricity board. Unlimited ESCROW capacity will be available. But, in practical situation, it is not going to happen. We have a structure of ESCROW. It is primarily the three tier ESCROW. It is principally the money from the dedicated circle that will flow through this ESCROW account. What happens is that the power company will first submit the bill to SEB and if the SEB does not pay, they go to the bank to get the money through NLC mechanism. In case the NLC mechanism also fails then only the ESCROW mechanism to the electricity board without any hindrance. Different model for collection exists where the dedicated circle, the geographical circle and the collection centres of the SEBs ensure collection. Then money from the collection centres to flow through ESCROW account.

There are some issues on the State Government guarantees. Some of the State governments have not enacted the Guarantee Act. Particularly IDBI has problems in Gujarat. Because, under their legislation there are certain issues regarding guaranteeing the commitment made to the IPPs under the Power Purchase Agreement for proper payment of bills.

The power project requires normally about 12-13, sometimes 15 years loan primarily; during the construction period ranging from 2-4 years plus 8-9 years of term loan. IDBI raised funds from the market on 5-7 years maturity. One suggestion that IDBI made to the government is that the long term funds are primarily from the Provident Funds. We have suggested to the government, that these Provident Funds should be allowed to be invested in a larger measure in the institutional bonds to raise our resources. Debt market also has lots of problems. Our debt market in the country is not deep like equity. A number of markets have equity shares and very few debt shares being transferred. We have problem on that account. So the problem here is that when securities, the assets of our own portfolio, IDBI is not able to securitise, because of the no depth in the debt market. RBI is seriously looking at this issue and will address it sooner.

Impact of fluctuation on the neptha based projects is also important. Assuming neptha at US $ 200. Now the prices have come down dramatically. But, assuming that this is the projection you will have a levelised tariff of 3.48% over a period of ten years. The fuel cost is about Rs. 1.94 per unit and the fixed cost is 1.49.

When there is FOREX fluctuation of 6% we find the levelised tariffs goes up to 3.92% so there is almost 12% increase in the tariff. Then we have 8% fluctuation, we are almost at 4.09 for the same fuel, with no other change, only for fluctuation. It does not take into account the domestic fluctuation of the Rupee.

How much the power projects are relying on the domestic financing can be found out by considering the first case. It is a project of Rs. 1,175 crores. A total institutional exposure is almost 75%. That means 30% debt and almost 70% by way of guarantee or by way of rupee loan. In the project two again equally 70%. So the reliance of domestic institutions is very high.

It is a typical example of cellular project how it works. It is again a infrastructure project and here you will find that FIs exposure as only 40%. In another case it is just 12%. You will find that lot of offshore debts are available for cellular projects, unlike there is non-resource funding. The reason is very simple. The reason is that is a market oriented. People are willing to take risk, but here it is not market determined. So, we have several constraints in the regulatory system. All the stopping fund flow unless FIs are willing to fund the project; unless there is guarantee from the institution or there is a counter-guarantee. The message is very clear. The faster pace of reforms process result in tremendous growth in the power sector.

ROLE POWER FINANCE CORPORATION

20

T.N. Thakur

Director (Technical), Power Finance Corporation, New Delhi

BACKGROUND

The Power Finance Corporation Limited (PFC), set-up in July 1986, as a Development Financial Institution dedicated to the Power Sector, provides funds to the various Power projects namely, renovation and modernisation (R&M), System Improvement, Urban Distribution, Transmission and Generation Projects. Besides, PFC as a DFI, also focuses on the institutional development of its borrowers.

BORROWERS

PFC's borrowers are primarily the State Electricity Boards, State Generation Corporations and State Electricity Departments. The lending portfolio has been broadened to include the joint, central and private sector power entities.

Eligibility Criteria for Borrowers

State Sectoral

State utilities to qualify for PFC loans, should comply with following conditions :

(a) Formulation of Operational and Financial Action Plan (OFAP)

(b) Achieving 3% Rate of Return on Net Fixed Assets; Condition relaxed for R&M, Capacitor, Environmental Upgradation, Studies and Consultancy Schemes

(c) Timely Submission of Annual Accounts

(d) No Defaults in repayments

(e) The amounts of financial assistance requested should be within the exposure limit.

The status of eligible and non-eligible borrowers as on 31-03-1998 is detailed in Annexure - A.

Private Sector

(a) The Borrower shall abide by Government of India rules and regulations governing private sector participation.

(b) Borrower should not have accumulated losses in immediately preceding three years and should not be in default to any of its lenders and annual accounts should indicate satisfactory financial performance.

(c) In case of new power sector entities, financial operational performance (track record) of the promoters for the past three years would be seen.

(d) In case of the existing power sector entities, the project should be compatible with their existing expansion plans, etc.

(e) Debt Service Coverage Ratio (DSCR) should not be less than the acceptable norm and Debt Equity Ratio should be in accordance with Government of India guidelines.

Lending Criteria

PFC assists only such projects which meet the following criteria:

— are economically justified—Minimum 12% EIRR

— are technically sound;

— technical solutions proposed are least-cost;

— all inputs and linkages are available;

— solutions proposed should meet Government of India or State environmental impact standards, whichever are more stringent; and

— schemes should have all clearance required by the State and Central agencies.

Priorities

As a Development Bank, PFC plays a very important role in guiding the power sector to an optimal path. The type of schemes identified for providing financial assistance on priority basis are

(i) Capacitors and System Improvement, Meters etc.

(ii) R&M of Generation

(iii) R&M of Transmission and UD Improvement

(iv) On-going Generation

(v) Completion of Missing Transmission Links

(vi) Urban Distributions and Transmission upto 132 KV

(vii) EHV Transmission above 132 KV

(viii) Generation—Small hydro

(ix) Generation—Large hydro

(x) Generation—Thermal

The extent of financing by the Corporation for various power projects, as percentage of the project cost, is given below :

Sl. No.	*Scheme*	*% of Financing*	
		State Utilities/ Municipal-run bodies	*Pvt. Sector*
1.	Pre-investment Studies	100%	50%
2.	Capacitors	80%	50%
3.	Renovation and Modernisation	70%	50%
4.	Transmission and Distribution	60%	50%
5.	Generation		
	— Small Hydro (Upto 15 MW)	50%	50%
	— Large Hydro	35%	25%
	— Thermal	30%	20%

INSTITUTIONAL DEVELOPMENT

Consistent with its envisaged role as a Development Financial Institution, PFC continues to attach high priority and importance to qualitative improvement in the functioning of SEBs/SGCs in managerial, technical and financial areas through formulation and implementation of Operational and Financial Action Plans (OFAPs) for its borrowers.

These Plans are prepared in consultation and agreement with the Utility and the State Governmnet concerned. Besides disseminated vital information through regular conduct of workshops/seminars, PFC also organises customised training for Power Sector Personnel in India and abroad.

Structural Reform-Initiatives

The Corporation has taken an initiative to provide technical/financial assistance to the State Governments/Power Utilities for structural reforms of the State Power Sector. A Reform Group has been constituted in the Institutional Development Division to guide and assist the State Governments in the formulation of restructuring programmes and for working out a suitable model most relevant to a particular State. Presently, interactions have been initiated with State Government/SEBs of Assam, Meghalaya, J&K, HP, West Bengal, Tamil Nadu, Kerala, Maharashtra, Goa and Punjab. Governments of Assam, West Bengal, J&K, Goa and Punjab have already communicated their willingness and commitment to restructure the State Power Sector with the assistance of PFC.

RESULTS OF OPERATIONS

Starting from a small beginning towards the end of 1987-88, PFC has made rapid progress. The total financial assistance (term loan, lease finance and bill discounting) sanctioned from its inception upto March, 1998 is Rs. 14069 crores with total disbursement aggregating over Rs. 9891 crores. Programme-wise details are given as under :

(Cumulative upto 31-03-98)
(Rs. in crores)

Programme	*Amount*	
	Sanctioned	*Disbursement*
R&M and R&U	1269	683
Capacitors	433	284
Transmission	4116	3161
Urban Distribution	1512	1099
Generation (incl. Lease Finance)	6532	4571
Studies/Consultancy	157	43
Bill Discounting	50	50
TOTAL	**14069**	**9891**
Working Capital Loan	50	50

Resources (As on 31-03-1998)

		Amount
1.	Equity from Government of India	1030
2.	Bonds	2437
3.	Foreign Currency Loans	1298
4.	Government of India Loans	1282
5.	Term Loan (SBI)	90
		6137

Operations

During the year 1997-98, the Corporation sanctioned Rs. 2922 crores towards term loan, leasing finance and bill discounting. The corresponding figure of sanction in 1996-97 was Rs. 2114 crores. Thus, there is increase of 38% in sanction in 1997-98 over previous year.

The disbursement during 1997-98 was Rs. 2019 crores as against Rs. 1552 crores in 1996-97. Thus there is increase in disbursement by 30% over last year.

PFC has shown remarkable progress in the areas of Resource Mobilisation as well. During the year 1997-98, the Corporation has successfully raised US $100 million at most competitive rates raised by any Indian Corporate for the tenure. PFC also raised Rs. 175 crores of tax-free bonds at a low rate of 8.85% p.a. and had obtained Rs. 90 crores term loan from SBI at prime lending rate (floating rate). During the year 1996-97, the Tax-Free Bonds Issue launched by the Corporation on private placement basis received an overwhelming response from the Investors—both institutional and individuals. The Corporation in its first venture into the commercial debt market since 1990 successfully secured a $ 75 million loan without a Government guarantee and at extremely competitive interest rates of 50 basis points over LIBOR. The rate negotiated by PFC was lowest so far on loan of 5 years average life.

The details on disbursement made and resources raised during the year 1997-98 are given below :

Resources raised during	Rs. in crores
1997-98	
Government of India Loans (WB/ADB)	382
ECBs	378
Bonds	175
Loan from SBI	90
Realisations	473
Internal Resources	453
TOTAL	**1911**

Disbursements	*Rs. in Crores*
During 1997-98	
Scheme	
R&M	92
Transmission	568
Urban Distribution	206
Shunt Capacitors	37
Generation	1063
Studies/Consultancy	23
Bills Discounted	30
TOTAL	**2019**
Working Capital	50

PROFIT RECORD

Year	*Net Profit after Tax (Rs. in Crs)*
1988-89	15
1989-90	30
1990-91	73
1991-92	108
1992-93	143
1993-94	148
1994-95	156
1995-96	211
1996-97	240
1997-98	481

DIVIDEND RECORD

PFC has been a profit making enterprise right since its inception. The dividend declared in the last three years is as follows :

Year	Amount (1997-98)
1994-95	30
1995-96	40
1996-97	48
1997-98	15 (Interim Dividend)

MOU RANKING

PFC has been consistently ranked 'Excellent' for its performance against MOU targets for three years beginning 1993-94, ever since it started signing MOU. For the year 1997-98, the Corporation is expected to be rated as 'Excellent' based on its performance. The performance and the composite score based on provisional data has been submitted to Department of Public Enterprises. Their formal communication of the rating is awaited.

MANPOWER

PFC is a lean organisation (total employees—245) endowed with highly qualified professional manpower. During the year 1997-98 the per employee disbursement touched Rs. 9 crores which rates high amongst the FIs.

FUTURE PERSPECTIVE

In the years a head, PFC's efforts will be directed towards further consolidating its position as the premier Development Financial Institution for the power sector. Towards this end, PFC with the active support of Government of India, would strive to emerge as the nodal agency through which all credits to power sector from bilateral/multilateral sources, would be channelled. PFC would also endeavour to facilitate structural reforms in the State Power Sector to make it commercially viable and efficient.

As against a disbursement of Rs. 1552 crores achieved in the terminal year of the Eighth Plan, PFC envisages to reach a disbursement level of about Rs. 5000 crores annually by the end of Ninth Plan.

Besides providing term finance to its borrowers, PFC has added lease finance; bill discounting; bill rediscounting; guarantee services; and consortium/syndicate financing to its portfolio. The range of financial services would be progressively broadened in future to include fund based services such as supplier's credit, bridge finance, equity participation, etc. and nonfund based services such as project consultancy, financial advisory services.

ANNEXURE 1

WEC ENERGY FACILITATION PROGRAMME (EFP) REQUESTS FOR EFP SERVICES FROM AFRICA AND SOUTH ASIA

Requests from African Member Committees.

Benin	:	Advice on Energy Policy and invitation to make presentation on applying actions to address Benin's energy issues.
Guinea	:	Advice on promoting private investment and risk analysis.
Democratic Republic of Congo	:	Advice on privatization of electricity distribution.
Niger	:	Promotion of renewable energy projects.
Lesotho	:	Advice on how to prepare bankable energy projects.
Côte d'Ivoire	:	Advice on policy and operations in petroleum sector.
Senegal	:	Planning for sustainable energy development into the next millennium.

South-Africa EFP requests include :

- Preparing a dossier of bankable energy projects integrating with rational energy policies
- Promotion of renewable energy investment
- Integration of energy policies in the Southern African region and advice on energy policies.

- Dissemination of best practices in policy and operations.
- Facilitating energy efficiency by integrating SADS energy efficiency policies.
- Facilitating creation of Southern Africa Energy Aid Programme to provide assistance to energy activities in Southern Africa to mobilize energy investment
- Rationalization of refineries in SADC region to minimize costs.

Swaziland : Advice on issues of energy sector restructuring.

Zimbabwe : Promoting renewable energy projects to investors and equipment suppliers.

UNECA/OAU : Support for establishing the African Energy Commission.

Requests for EFP services from South-Asia Member Committees

India : EFP services requests include:

— Restructuring and promoting private investment in energy.

— Addressing policy and implementation issues and preparing action plan.

— Integrating the national policy and implementation issues into regional programme.

— Preparing dossier of bankable investment projects at the national and regional level.

Nepal : Review of existing acts, regulations and guidelines, formulation of power purchase agreements and sharing of downstream benefits from hydro resources.

SriLanka : Natural gas utilization and promoting investment in renewable energy development.

DISCUSSION

Question: Dr. R. Vedavalli

My question is how has the high domestic interest rate affected the cost of power and its borrowing in the domestic markets. This is for Mr. Jain.

Answer: Mr. P. C. Jain

Basic premise is that the domestic loans are costlier than the foreign loans. Yes, on the face of it looks like that.

Today, if you go for the Yen borrowing, it is around 3-4%; for the Dollar it is 8-9%; and if you borrow domestically it will be 13, 14, 15%. But we have analysed it and we find that facts are just the reverse in order. In our past borrowings, our costliest borrowing is the Yen borrowing. The next costly is the Dollar brrowing, cheapest is the Rupee borrowing. So, I think the basic premise, that the Rupee borrowing is costlier is not correct.

Question : Mr. Ali Attiga

I was happy to hear the hydro-carbon sector mentioned today because until that happens I wondered whether energy meant in this Seminar was synonimous with electricity and coal. I was impressed early by what India had done in the hydro-carbon sector. As I mentioned, 11 years ago, I came on extensive visit and I saw Mumbai High, the activities you were doing and I had expected that India by now would be forming companies that would explore for oil and gas both. I am right to say, you have no financing problems in hydro-carbon sector. You are able to mobilise resources as you need but how is the international oil price affecting you. The prevailing oil price and in the next 5 years is going to be low. Then, how much is the government taxing the oil products to the final consumers? I did not hear the question of oil price at all mentioned in your presentation on financing. I ask if it does not have any relationship?

Answer : Mr. A.K.T. Chari

You are absolutely right that oil pricing have a major impact on financing situation. Obviously, I did not go into all aspects, which really gets into financing of the oil sector and is not possible to cover in the given time. Some of the issues are as follows:

a) In the present situation of National Oil companies we do not envisage much problem in generating financing in India or abroad for our projects;

b) The reform process, on pricing, are now linked to import priority. Most of them have been de-regulated in terms of number of products. 30% have been de-regulated where prices are linked to international market and rest of the 70% government is committed to de-regulate over the next three and a half years;

c) The international prices are so low at the present moment and not ony low, there are certain foreseen aberrations in prices on different products which should be avoided. That is resulting in margin squeeze for the companies. It has an adverse impact in a developing country where the capacity is a part of infrastructure. At the moment we have enough projects which will meet our capacity and the refining should more than double in the next 8-10 years, subject to the GDP growth, remaining as till now, i.e. 6-7%, over the next 10 years period.

Finally, we are looking very actively at opportunities abroad for equity as well as for joint ventures. From IOC we are in close touch with ONGC dealers and with others to put our stake abroad, Central Asia as well as South Asia. We are very keen to put our investments both in finance and technology into these areas.

Question from the Floor

1. I have two questions basically from Mr. Chari. The whole idea of opening up the power sector 1992 was to get additionality of resources from the international market. Your presentation, if I saw it correctly, shows already claim 70% of debt of total funding seems to be apparently coming from the Indian financial institutions. Keeping the sanctions in mind, could you please guide us and tell us why this is happening; and is there any situation where international lenders are having a problem of lending projects to India?

2. Was liquid fuel foreign exchange sensitive. Whether LNG is foreign exchange sensitive or not in your opinion.

Answer : Mr. A.K.T. Chari

Your last question first. LNG is foreign exchange sensitive because it is going to be imported, as we do not have any resource now. LNG prices are cheaper. We have not gone into details, because there are no projects on LNG use currently. But, based on a number of operations using LNG which we have seen that it is felt that LNG will be cheaper in power generation than using naptha. Of course it depends on your transportation cost.

Answering your first question I have to say that regarding 70%, it includes 40% debt, plus 30% guarantees. That is the standard practice. Usually we do 40-30% debt ratio. 30% equity is coming from the sponsors and 40% debt sealing is there and balance 30% is coming as foreign debt. The foreign debt and part of the guarantee is coming from the financial institutions and banks, so that is why the exposure of 70% It covers not only the fund based but also non-fund based guarantee.

Another question is as to how the foreign sponsor look at it? Even today we have a number of visitors coming to IDBI and looking for cooperation from the IPPs and funding new power projects. So, look at many of the IPPs out of the 30% equity, very few have gone for public issue so far, but are inquiring through the foreign investors plus their own funds. That means there is tremendous appetite for power projects. Everyone is looking towards the direction of the reforms. That is going to be a very crucial factor now in future of the public sector.

Question : K.N. Sinha

Jain mentioned about reported financing and said that when we go in for balance-sheet financing the lending cost comes down. My first worry is that taking a clue from that of financial institution, you feel comfortable that the projects are close to balance-sheet and if so would you recommend that the government should come out with false statement on the subject and make conditions that financing in the power sector would be made only on the balance-sheet. Will it add more confusion to what already exists/ Secondly, is it hitherto the Indian financial institutions were not charging the fees, commitment charges, banking charges and what not. But, once we got exposed to the international market, the Indian financial institutions are also raising those charges in the loan package etc. Be frank. Do you think that is required and a necessary thing, because it is raising the cost by 1.5 to 1.75%?

Answer : Mr. A.K.T. Chari

Currently the institutional charge is at a uniform rate of 1.05%. It is an international practice. We do not charge any other management fee or anything. This 1.05% is not a recent phenomenon. We are looking at the fund raising problem. We raise the funds and that are sometimes kept unutilised. Looking at those kind of factors that it is a reasonable amount. Coming to the first question, balance-sheet financing. Probably I answer this way. In the project it is not new to us. The reforms started only in 1992-93. Even earlier, we have financed number of projects including BSES,TATA Electricity etc. This had certain peculiarities, they had their licensed area. Therefore, market was not the problem, but still the tariff regulation was with the State Government. But States guaranteed minimum return; there was no subsidy. But, there was always as some kind of one paise, two paise per unit as subscription was coming.

Discussion among the Penalists

But ultimately marketing was not a problem. We are taking recourse to the project not to the companies' balance-sheet.

Comment: Mr. P.M. Chacko

I have been concerned about the volume of funding involved. Inspite of all that has been said reassuring about the financing, particularly the domestic capital size, there was 10,000 MW, may be around Rs. 40,000 crore. Well at the end the requirement may be Rs. 70,000 crores anually. About 75% comes from the domestic capital market. I am astonished what you have said yesterday, if we go further for the larger money abroad we are getting into the debt trap and something like that. So this much quantum of money had to be there. Initially we had a rate of growth of 12.2% . From that rate of growth we have slumped to around 4% growth. Let us hope when there is crisis that will be the way out.

PART-VI
ACCESSING INTERNATIONAL FINANCIAL MARKET

FINANCING OF RENEWABLE ENERGY PROJECTS

22

H.S. Rawat
Manager, I R E D A

IREDA was set up in 1987 to support financing of the renewable energy projects in India on soft term loans, which at present varies from 2.5% to 14%. Within a short span of six years, IREDA achieved success in getting international funding. World Bank extended a line of credit of US $ 195 million in the year 1993-94. More than 95% amount has already been rolled. This covers solar photovoltaic and small hydro and wind sectors. In the second line of credit, LADB offered US $ 120 million. It covers solar, thermal and wind power generation. The World Bank offered another line of credit of US $ 173 million. It will cover smaller hydro and energy efficiency projects.

Though IREDA was set up in 1987, its work actually started in 1991. So far, we have sanctioned 963 projects, covering Rs. 1,500 crores. It is not a matter of fact that renewable energy projects may be small but in the gamut of the development skill in social sector upgradation and other consolidation approach, renewable energy has played a vital role.

Recently, IREDA exercised some innovative models in Sunderbans—an inaccessible area. The entire Sugar Island is converted into Solar Island. Ministry has given a very good impact over the solar photovoltaic programme; and small hydro programme is also going at a very good pace. The World Bank has given an excellent rating to IREDA. IREDA expects that by 2002 it can access the international fund to the tune of Rs. 3,000 crores. IREDA's contribution to the Renewable Energy is 1.2% of the entire power sector programme and I thought that whenever you are giving some sort of good idea to access international fund or domestic fund or multilateral or some other EUROBOND one must consider the role of renewable energy sector.

The tragedy of 50 years is that more than 87,000 MW came into the grid. But very little has been diverted to the social group. Nobody has been thought of it. I am from a village and I have seen the entire country's development. For the last 50 years, there is no electricity in my village. Only renewable energy projects can go to far-flung areas. If you are having that diction, if you have that mission and I should achieve the self-sufficiency over the programme you to change your directions as the situation demands. Because 1,000 MW era is over, it is becoming very difficult to get the money from the international financial corporations or any other funding agency. But cannot sit idle.

Being a small company, we have generated 963 projects in six years to fund the entire capacity addition of 400 MW. Although 400 MW is very small, IREDA is classified under socially sector financial organisations. So, we are having exercising attitude on this. Although IREDA is very small in comparison to IDBI, IFCI, our approach is totally missionary. We are trying to push something in the social sector. Being a small 1 MW, 2 MW, 20 MW, but that is going to serve the entire DSM programme. IREDA is facing the problem of quantum approach. In future, many power purchase agreements will come—including the mega projects, national grid and other local grids has succeeded in achieving 95% rural electrification. I am ashamed to say in the last 50 years, India has got only 140 MW in small hydro, whereas China has got 56,000 MW in 50 years span. If we neglect a small candle not to be commissioned in the hut of the rural areas, it is immaterial. Electricity is not an allied property. It is not a property of big financial institutions. It is intention, your approach, your innovative financial engineering programme and to get a success in this entire programme, depends on how you are motivated.

In Vivekananda's speech, he used to say, "If I get 100 young boys, I can move the entire world". Like this, if you see in the real sense, we have neglected the social sector for a long time. We have to concentrate on this and I think the problem in getting access to the international funding will be solved. We are very confident that our IREDA's staff portfolio is only 60 officers and the entire total strength is 120. 120 officials can mobilise 975 projects in India. Why not IDBI, IFCI? I think there is a difference in approach.

PFC'S EFFORTS AT ACCESSING INTERNATIONAL CAPITAL MARKET

23

T.N. Thakur,
Director, Finance, Power Finance Corporation

Power sector needs longer tenure loans and we deal moratorium for six months after the date of next commencement, this is 3-4 years and then there is the repayment period of 8-10 years. So, average gap comes to 13-14 years. ADB and World Bank have offered loans of longer tenure; and therefore, asset liability mismatch was not there. Still PFC were getting concerned about the mismatch which might appear in over a period of time; and therefore, PFC started looking for off-shore borrowing for the sake of long term. Thrust was on syndicated loan of 5 years average maturity and observed that the loan was very well received. PFC went in for a small amount of US $ 50 m, six syndicated loans. At the first instance PFC got US $ 80 m. Encouraged by the first offshore market response to PFC, it was seen that the investors perceive the power sector as a great potential in India; and they were also looking at PFC as a kind of window for power sector funding. Encouraged by that the next issue PFC wanted to raise was the 12-year maturity EURONET Bond. The US and the European markets were tapped for that. It was again US $ 100 million on a fine rate of seven and half percent was the finest rate that any company was getting at that point of time from this part of the world. After PFC completed that issue; effort was made in the South Asian markets. PFC's further programme for going for longer tenure, debt from offshore market was kept on hold for some time. Only when the market situation started improving, PFC again went for the issue.

PFC has resource mobilisation plan of about Rs. 2,000 crores during the current financial year (1998-99); and that will go on increasing during the whole plan period. PFC has resource mobilisation plan of Rs. 17,000 crores. This we have to raise from the domestic market as well as from the offshore market. Exact mix as to how much is to be raised from which market is not predetermined and that depends on the market conditions. As of today, there is lot of liquidity in the Indian market. The investors in the Indian market are looking for real good investment. Actually PFC have negotiated some loans from the banks in India at fine lending rates. PFRC is taking those loans because it is observed that in the present day scenario the loans are working out to be cheaper than raising bonds in the market or offshore loans. Most of debts are available. Experts feel that the back up security arrangement is alternative to counter guarantee. Those things are only fall back arrangements. What is important is to look at the cash flows in the system.

If the system has cash flows for supporting debt servicing, then fall back arrangements will take care. That was quite amply demonstrated when PFC went in for the debt issues. We have gone unsecured, without any guarantee, without any letters of credit and yet the investors have so liberally invested in PFC bonds issue. They did not ask for counter guarantee; they did not ask for the guarantee from the Government of India; and they did not ask for security from PFC. So, what is important is that we are able to demonstrate to them that PFC has sufficient cash flow to take care of the entire debt servicing during the tenure of the debt. The tenure was decided on the basis of asset liability matching, it could not be determined on the basis of market conditions. And, when PFC finds that loan from banks are cheaper then the bonds it is preferring that option. PFC's total debt mobilisation so far is Rs. 2030 crore of tax free bonds; Rs. 3,000 crores of taxable bonds; and Rs. 2,000 crores of SLRS. PFC also has a small fixed deposit scheme.

PFC is expanding and diversifying its investor base so that the reason for PFC going to US market, following the Rule 144A. We also planned to have some public issues. PFC is also trying for standing arrangement for medium term loan, so that it can access the market at any point of time. PFC has been in dialogue with some foreign banks and financial institutions in regard to structuring certain kinds of finances for the power projects. As a result of the financial engineering undertaken by PFC, it was possible to reduce the cost of particular project by about 10% when PFC organised foreign debt for them instead of their own. PFC has EURONET bonds 31% of the total.

PFC has the mix of different currencies in our basket of foreign currency loans. PFC has established umbrella agreements. The framework agreement with some of the banks and some of the export agencies having already been tied up on the terms and conditions even the pricing, but at the cost and no commitment on the part of PFC. The commitment and cost will start only when there is an agreement for a particular power project and for definite time. There was lot of interest created amongst the banks and other financial institutions to have this kind of arrangement with Power Finance Corporation. If the investor or borrower come to PFC, it can organise some kind of loan.

The syndicated loans and structured dealings can still take place at competitive prices because the people who participated in syndicated loans and structured deals, do that with due deligence. But as far as Bonds issue and GDR issues are concerned, PFC will wait for the market perceptions improve.

ADB'S APPROACH TO ENERGY SECTOR FINANCING

24

Dr. C.S. Chung,
Senior Energy Adviser, ADB

During last few months, ADB has imposed sanctions on India. I would like to clarify our position to India. Actually, ADB has never imposed sanctions on India. That is why naturally there is no such thing as lifting sanctions. ADB operations in India are going on as usual. So, our cooperation in the energy sector with PFC, NTPC, ONGC and also IREDA will continue. But only the ADB project, processing and approval, procedures will be followed. If any project has been delayed, it is not because of sanctions. It is partly because of project's specific issues. But hopefully that will be resolved in due course.

Actually the resources of multi-national institutions such as World Bank and ADB are comparatively limited. So, the World Bank and the ADB are the main sources to provide these funds for the energy sector. But, now I would not say anything more, because private sector has grown very much and also your fund requirement in the energy sector is huge like your power sector. How much MW you have to produce or to deal in the next decade. Huge amount is required to produce 70,000 MW to 85,000 MW. So what we are doing is that the generating sector of the power sector will be financed by the private sector and so expect power sector to play a vital role. Dealing with the transmission and distribution, requires involvement of SEBs, State government and the Central organisations like PFC and NTPC. So ADB lending will be some sort of currency related lending. So in such cases ADB would not like to make loans to SEBs, who are financially quite weak. ADB like to make these SEBs financially viable organisations, so that the IPPs and the private sector will feel more comfortable.

ADB have already provided huge amount of money in the hydro-carbon sector in India during last 5 to 6 years about US \$ 1.3 billion was given to India for hydro-carbon sector. They are now in a good position, ONGC, that took money from ADB and is a good result oriented organisation. So ADB's operation in hydro-carbon sector is very successful, and their funding demand is also huge. So our role will be limited in the energy sector in future.

ONGC'S APPROACH TO ENERGY SECTOR FINANCING

25

Dr. A. Bhattacharyya
Economic Adviser, ONGC

As regards financing, upto 75% is mobilised through allocation of public funds or equity. Only 25% is to be mobilised from the market sources. Out of this 5% on an average by syndicating loans and another 5% from the technical assistance and soft loans from the World Bank and ADB. I am talking about those projects which have been completed.

ADB has, in the past supported gas fired reduction projects and another re-engineering project. Similar assistance can be expected in future also. ADB assistance may be available in future also. Another 5% of Rs. 6,000 - 7,000 crores funding is arranged through the suppliers' credit of transnational companies and also we are exploring the routes of ECB and ECA because we also get deemed export status; which is also extended to ONGC for imported capital goods for exploitation activities. Another 5% is already sources from different bonds like Samurai and Exim. So, I am coming back again to the Philosophy of ONGC's financing of different projects. ONGC is no longer only a production company. It has plans expand into a US $ 10 billion company by 2010-11. Therefore, ONGC is looking for a different sourcing mechanism and focus on the financing of different firmed up projects internationally. ONGC has in Bangladesh a US $ 50 billion project for drilling activities. Similarly, other projects are in Kazhakistan, Tunisia, Indonesia and Iran and other places. Already, Petro-Vietnam project is on by our ONGC Videsh Ltd., a subsidy of ONGC and those operations are being conducted through mostly internal resources to some extent. As these are large investment projects, it will be necessary to depend on commercial borrowings and also other sources of the debt through different instruments. This is the ONGC approach to project funding.

If at all, there is a capital shortage, it is in local markets, but not in the global markets.

PART-VII

RESTRUCTURING OF STATE ELECTRICITY BOARDS AND ENERGY UTILITIES

RESTRUCTURING OF STATE ELECTRICITY BOARDS AND ENERGY UTILITIES

26

R.N. Srivastava
Chairman, Central Electricity Authority, New Delhi

INTRODUCTION

India, despite having increased its installed generating capacity manifolds continues to face persistent power shortages primarily due to demand growth outpacing the growth in generation. The condition has reached a very critical stage in view of the crucial role the electricity plays in the economic development of the country. With a view to sustain the fast growth rate and constraints of matching fund availability, a conscious decision was taken in 1991 to open up this important infrastructure sector for private participation. Though there has been good response in terms of the interest shown by various agencies, the results have not been as encouraging as were expected.

It was envisaged to install additional capacity of 30,000 MW in the Eighth Five Year Plan. But the capacity addition actually achieved was around 16,500 MW. The effect of shortfall in power capacity addition during the Eighth Plan has resulted in peaking shortage, of more than 18%. As per the projections made by the 15th Electric Power Survey, the peak demand requirements are likely to increase to around 95,700 MW by the end of Ninth Five Year Plan (2001-02). The capacity addition envisaged during the Ninth Five Year Plan is around 40,225 MW. The fund requirement based on this generation capacity addition would be enormous say around Rs. 1,20,000 crores. Further, the fund requirements for transmission and distribution sector would be around Rs. 44,000 crores. The capacity addition envisaged during the Tenth Five Year Plan is 22,200 MW. The bulk of share for this capacity addition is expected from private sector. It is, therefore, essential to provide suitable regulatory mechanism and institutional set-up for efficient management of power sector of such a large magnitude, particularly due to coexistence of State and private sector.

REFORMS-OBJECTIVES

The objective of any reform/restructuring programme in the power sector is to achieve efficiency and to provide quality supply/service to consumers at an affordable price. The Government's role should be restricted to that of a facilitator and promoter and for development of healthy and transparent policies. The emphasis on structural, institutional and regulatory reforms should aim at independent and effective enforcement mechanism, transparency, attainability and easy access to information data for all concerned. The reforms also need to provide for a proper coordination between the organisational set-ups at the State as well as Centre level for framing future regulatory structure as any sudden shake-up may endanger the continuity of development in the country. The essence of any reform is to increase the efficiency by reducing the cost of generation and system inefficiency and to promote competition with minimum regulations and also safeguard the interest of various players in the field.

EXISTING LEGISLATION

Under the Constitution of India, Electricity is a concurrent subject. Thus, the Centre as well as the State Governments have authority to enact legislation in regard to the power sector. The power sector entities in India are governed by the following statutes:

— Indian Electricity Act (IEA) 1910

— Electricity (Supply) Act (ESA) 1948

— Indian Electricity Rules (IER) 1956

Indian Electricity Act 1910

The intent of Indian Electricity Act 1910 is to impose certain duties and liabilities on licensees and other persons who supply electrical energy in an area for which a licence or permission is granted. The power granted to licensees or other persons are conferred for the purpose of enabling them to construct their works, plants, service mains etc. and to maintain them. The duties are imposed on them for the safety of public or individuals who purchase energy from them.

Electricity (Supply) Act 1948

The main objective of Electricity Supply Act 1948 is to promote coordinated development of electricity in India and to extend the benefits of electricity to semi-urban and rural areas in most efficient and economical manner. The Act institutionalised the single State-owned integrated monopoly organisation namely, SEB for generation, transmission and distribution of electricity for the entire area of the State.

The Electricity Supply Act 1948 also provided for creation of Central Electricity Authority (CEA) as an advisory body to the Ministry of Power. The powers and functions of CEA include planning and development of the power sector policy for the country arbitration in the matters arising between State for projects

The Central Government can direct CEA; and State Governments can direct SEBs in policy matters.

Safety Aspects

The Centre and State Governments have the right to appoint Electrical Inspector as per Section 36 of Indian Electricity Act 1910. The office of Electrical Inspectorate is responsible for ensuring the safety of electrical installations and equipment in power system. The Act, however, have not ensured adequately judicial independence to the electrical inspectors as it continue to be total subordinate to the Central and State Governments which have overriding powers to exempt certain provisions of the Indian Electricity Rules.

(a) Rules/procedures are not backed up by enforcement provisions

(b) Principle of tariff fixation at consumer level not clearly stipulated

(c) SEBs being responsible for generation, transmission and distribution cannot regulate themselves.

Thus, there arises the need for separate regulators.

NEW DEVELOPMENTS

As the existing legislation (ESA and IEA) do not permit regulatory/structural changes, the power sector reforms initiated by the States envisaged creation of an independent regulatory commission and introduction of more players in generation and distribution through a process of unbundling. The typical model of restructuring contemplate dismantling of the present monolithic structure of the SEBs, separation of functions of electricity generation, electricity transmission and electricity distribution. If we examine the managerial and technological capability required in all these three functions, there is not much in common. While the generation of power is an activity which needs tying up of number of inputs to produce electricity, the transmission industry is a service for the purpose of providing network for transport of power produced by generating companies. On the other hand, the distribution activity is related to a service for supply of power to consumer which needs lot of interaction with the consumer for metering of energy consumed, billing and revenue collection etc. All over the world, therefore, a view has developed to unbundle the total sector. It means dismantling of vertically integrated monopolies of the State and public utilities. The electricity transmission system not only provides a network to transport and maintain power with reliability but also works as a market mechanism where all generators connected to it can supply power in competition to each other to various distribution systems connected to the transmission network.

The multiple entities proposed to be created in a post-restructuring scenario would need regulatory mechanism. To create level playing field between the SEB (or its successor entity) and private licensees, regulatory powers of SEBs would have to be withdrawn. The emphasis of regulation would have to be price, quality and reliability of power supply to the end user. Based on these principles various State Governments have prepared State Regulatory Bills which enable the creation of the Regulatory Commissions in the States. These Bills also provide for changes in the current legislation in order to allow the existing and new entities in the power sector in the States to function independently.

The Government of Orissa was the first to initiate the reform process through Orissa Electricity Reforms Act 1995. Under this Act, Orissa Generating Company having thermal units under it, Orissa Hydro Electric Generation Company, Orissa Grid Company and the Orissa Regulatory Electricity Commission have been formed. The initial reports of Orissa experiment seem to indicate that the results are positive. But we may have to wait for some more time before a final view is taken in the matter. The State has also initiated the second phase of reform by taking up privatisation of distribution system. The Electricity Reforms Bills of other States such as Haryana, Andhra Pradesh, Karnataka, Madhya Pradesh, etc., have been developed on the similar lines to that of Orissa Electricity Regulatory Act. The various issues which need to be addressed in these Reform Bills are as follows:

(i) There should be a time limit for the Government to select the members of Regulatory Commission.

(ii) Licensing Authority should be vested with Regulatory Commission and not with the State Governments.

(iii) Local Municipal Undertakings engaged in the distribution of electricity should be brought under the purview of the Regulatory Commission.

(iv) Any tariff subsidy should be made good to licensees on a regular basis. The payment should be covered by a suitable security mechanism acceptable to the Regulatory Commission.

(v) Any failure to comply with the quality standards by the suppliers and licensees should invite penalty provision to be enforced by the regulatory commission.

(vi) All unmetered connection may be deemed illegal and metering of all supplies should be made mandatory.

STRUCTURAL ISSUES

At the initial stages of power sector reforms it is essential to consider following aspects :

(a) Extent of Competition

(b) Extent of regulation

(c) Phasing of reforms/restructuring

The competition in the power sector is possible at the following levels:

Competition at entry level can be achieved through selection of players by competitive bidding (either in generation or in distribution). In the next stage, competition in bulk supply is possible where bulk consumers and distribution companies enter into direct agreements with generating companies without being forced to purchase power from a single agent. The competition in retail supply can be initiated when retail consumers have the freedom to choose their supplier.

The activities of generation and distribution can be considered as separate entities in order to facilitate introduction of competition and to make these physically manageable. The initial power industry structure under the refrom/restructuring programme would resemble the single buyer model which is similar to the reforms undertaken by the Orissa State Government under Orissa Electricity Reforms Act. In this model, there would be a long-term power sales/purchase contracts between the generators and transmission companies and between transmission and the zonal distribution companies. Distribution companies would have service areas monopoly and individual consumers (either bulk or retail) would not have freedom to choose their suppliers. The competition would be there only in the selection of private developers/promoters. The regulator would play the role of efficiency improvement through performance benchmarks. The next stage in the industry structure would resemble the multi buyers model in which increased competition in supply would have to be introduced in a phased manner, so as to achieve maximum benefits for the consumers. Under this arrangement, there would be multiple generators selling power directly to the distribution companies through a long-term contract and to a pool through pool purchasing price. There would be open access to transmission/distribution networks and the users of transmission/distribution network service would have to operate under a Grid code for transmission/distribution to be approved by Regulatory Commission. In this scenario there would be greater degree of competition as the customers would have choice to procure power from more competitive suppliers.

CONCLUSION

The Indian power sector at present is passing through a complex phase and is charged with the necessity to unbundle SEBs and to impose a fresh legal framework by creating regulatory mechanism. Although it may look attractive, any dilution of uniform national policy would result in sub-optimal use of national power resources and may prove costly to the national economy. The model of reforms should emerge out of necessary to operate SEBs on commercial lines and aim at correcting shortcomings. The main agenda of reform should be the electricity accounting and commercialisation of SEBs so that over a period of time real competition and efficient operations would emerge. The careful consideration of all aspects is essential keeping in view the overall interest of the country.

RESTRUCTURING OF ORISSA STATE ELECTRICITY BOARD

27

R.V. Shahi
Chairman and Managing Director, BSES Limited

About some of the features of the CES Act. The 1910 Act was written to create license. The licensee was created because power supply was restricted to city areas only. The Act did not deal with the fixation of tariff. The municipality used to fix the tariff of the licensee. No principle was laid in the 1910 Act. Subsequently, the Electricity Act, 1948 was written. Principles of Tariff fixation were specified for the creature created by the 1910 Act. The Sixth Schedule was beautifully drafted. It dealt with every aspect. This Act also stated what the city power stations will not do. These had to be integrated with the States. The 1948 Act did not consider the situation of going outside a State. In 1964, the Power Ministers met and decided in favour of Regional Electricity Boards. The Resolution adopted in 1964 created Regional Electricity Boards. No further action was taken on the Resolution. In 1991, the government formalised this Resolution. Chronologically, the Electricity Act was written in 1910; Sixth Schedule added in 1948; REBs created in 1964; and regularised in 1991. Power Sector from 1964 to 1991 tolerated RLDC and REB. Rajyadyaksha Committee recommended Regional Electricity Authority. When REBs were created in 1989, it was felt that REBs were not good enough. A National Power Grid was recommended, and created in 1989. But, there was no instrument in the Act that legitimised the National Power Grid. A lot of juglary was done and it was legitimised. The Transmission Bill has been introduced in the Parliament on July 16, 1998. Perhaps the creature created in 1989 would get legitimised in 1989. In the Electricity Regulatory Commission Act, 1998 tariff reforms have also been added and the need for National Grid has been identified. Thereby, the entire function and regulations of the National Grid has been prescribed in the CERC Act. Authority has been delineated so that this National Grid can be effective.

When 100 MW of power was to be transmitted from Orissa to Kerala, it took one year to create the enabling instruments. With great deal of work done in CEA, this enabling instrument was created by an administrative mechanism. It is a tragedy that there is surplus power in the Eastern Region, but it cannot be moved out, because there are no enabling instruments. The new regulators will be able to write down the regulations on transfer of power. That will enable, to evacuate 3000 MW from the Eastern Region. Physical parameters are different, physical lines are different but the instrument of transfer is much more complex, software is much more complex than the hardware which has been created and I hope this will do on further.

The real problems of electricity industry in India has been the paying capacity, the credit worthiness of the SEBs. During the period 1992-97, the losses of electricity boards have increased from Rs. 4,000 crores per year to Rs. 10,000 crores per year. The tariff which used to cover 82% of the cost of supply in 1992-93 covers hardly 78% of the cost of supply in 1997-98. That being the case, it has been an uphill task to attract the investors to come in the generation projects. So much So, now there is a growing realisation amongst the developers as also the authorities those who are formulating policies, that perhaps the right end of privatisation of electricity industry would have been the distribution rather than generation. But, if we do not set right the commercial side of the total electricity industry, how is it that we are going to serve the massive capital investments which are necessary for this sector.

1,00,000 MW that we wish to create in the next ten years will attract 3,50,000 crores of investment in generation alone; and an equal investment in transmission and distribution and fuel sector. Therefore, distribution side of the business has to be set right. That was the essence of the Orissa electricity industry. They decided that the present form and the form of the electricity industry which was SEB, is also the pattern in the whole of the country. SEBs control generation, transmission and distribution barring Karnataka where we have Mr. Jairaj as the Managing Director of the Karnataka Power Corporation, where distribution and transmission is different; and generation is different. Barring that, in the entire country we have generation, transmission and distribution vested on the electricity boards. This gave rise to the thought that we should restructure the industry in Orissa State. Should we disintegrate what the consultants call "unbundle" the industry. Unbundle SEBs to create three entities—generation, transmission and distribution. That is how Orissa Electricity Reform Act was passed in January, 1996. Orissa did the preparation over a long period of time, discussing at various official levels and a large number of consultants on various aspects of industry. A legislative scheme of Orissa Electricity Bill was formulated, which got into an Act in 1996. Only then the Orissa Electricity Generation Company was formed. Hydroelectricity Generation is separate and Grid Corporation of Orissa is separate. Distribution was kept with Grid Corporation of Orissa in the interim period pending formation of four distribution companies for four zones. At that point of time, some specific action took place in the World Bank or other forums.

It was considered whether, as an interim measure, the Management Contract could be interjected as a prelude to privatisation of distribution. That is how sometime in 1995-96,

Orissa State Electricity Board and Government of Orissa decided that in the central zone (which consists of Bhubaneswar, Cuttack and Dhenkanal) a Management Contract may be tried on tender basis. Three/four companies participated which included Calcutta Electricity Company and Bombay Suburban Electric Supply Co., BSES was finally selected to be the Management contractor for the central zone. The expert opinion was that under Management Contract, BSES may have no motivation for investment, and SEB will have no control. Therefore, some officials were sent on deputation. But, due to some court cases, it was decided by the tribunal not to follow that route. Therefore, control over employee was not kept subsequently.

This became the first hurdle in the success of the Management Contract. In any case, Management Contract was not the lasting solution. It was only to be a prelude to privatisation of distribution.

At this point of time, Orissa has a separate generation company and it is in the process of divestment. Recently, Orissa Power Company divested or is divesting about 49% of its equity. The idea is that generation company would also be in the private management control. The Grid Corporation of Orissa will control the transmission system for the entire State for 220 KVA and above. Orissa will have four Distribution Companies which will be gradually privatised. Orissa has advertised for shortlisting of agencies which would be taking 51% of equity in the distribution companies. About a dozen companies have shown interest and the process of short-listing is in progress. By March 1999, Orissa will have four zonal distribution companies which will operate as Joint Venture companies, with the State Electricity Board of Orissa of Grid Corporation of Orissa, GRIDCO having 49% equity and the strategic joint venture partner having 51% equity. But, massive privatisation covering the whole State would be difficult. If it is proposed to cover rural areas, semi-urban area, urban areas and agricultural and industrial sector altogether, it will be a very difficult task. The Sixth Schedule has an extensive presentation, very extensive coverage. Electricity Act, 1910 and Electricity Supply Act, 1948 as they stand without any amendment, do not pose any problem to extend the coverage to the private sector. The Act allows coverage of every geographical area covered under the licensee's arrangements. It does not prohibit the State Electricity Regulatory Commission to extend the coverage of 3% distribution in the private sector to a larger percentage of distribution in the private sector.

The country would be better served if a target of 25% distribution in the private sector in the next five years in planned. Subsequently, 5 years can witness further 25% privatisation. These targets should cover rural areas also. Covering urban and rural areas at the same time will create some problems. So, the starting point can be urban areas.

EXPERIENCE OF SEPARATE POWER PRODUCING COMPANY

28

K. Jairaj
Chairman and Managing Director,
Karnataka Power Corporation Ltd.

The Karnataka Power Corporation Ltd., KPCL is a power generation company and is one of the earliest companies to be formed out of unbundling of State Electricity Board. In 1970, the State of Karnataka separated generation and distribution; and for generation KPCL was registered under the Companies Act. So, part of the reforms that are mentioned now in terms of corporatising the utilities and separating generation and distribution was already achieved in Karnataka, many years ago. The only customer that KPCL has is the Electricity Board. From that point of view, I would like to present the case for reforms.

There are actually three major reasons why today reforms are imperative.

(i) Financing;

(ii) Organisation; and

(iii) Managerial.

These three issues can be dealt as follows :

Financing

The total loss of Electricity Boards is about Rs. 10,000 crores. The principal reasons for the losses is disparity between the cost of power in distribution and the realisation in terms of that power. The gap between the demand and supply is of the order of 20-25%.

Average generation cost of 1996-97 was 186 paise per unit, whereas average realisation today is 149 paise per unit. So, the gap for every unit power distributed is approximately 36 paise or 20%. This is true in all States of India without exception. That is the principle reason why this sector is today on the verge of financial insolvency. For instance, the receivable for Karnataka Power Corporation from the Karnataka Electricity Board runs into about 5 month's billing. Similarly, the receivable of NTPC, which a principle generating utility, runs into over Rs. 6,000 crores. The gap between cost of generation and distribution is one reason calling for reforms and tariffs.

The cost of power generation also marks many inefficiencies. Official, the T&D losses are quoted at 18-25%. The real figure may be much higher. Therefore, one of the ways which we have to make frontal attack on the matter of transmission and distribution losses and improve efficiency is by introducing reforms in this sector.

The third important financial issue is that the public investment is inadequate for development of generation or distribution sectors. Private investment has to be attracted. In order to attract private investment, a security mechanism to sustain that investment and minimise the risk is required. None of our State Electricity Boards have ESCROW capabilities to the extent required to sustain huge generation capacity. If the electricity boards are really not providing for ESCROWS, it is clear that we cannot match it in terms of the actual revenue stream that they have. One important point is that there is a critical financial disturbing force in our system and that is the percentage of paying customers is declining. The reference is to high tension industrial customeRs. Certainly, I have the figure which show that in the country as a whole and all States in particular, the consumption of power of HT customers is declining over the years because they have developed captive power generation. Take State of Karnataka, the number of HT consumers who consumed 35% of the power distributed have now reduced it to 20-27%. It is true of Andhra Pradesh, and Tamil Nadu. Similar pattern is seen on an All India basis.

Organisation and Management

The major problem is of ownership. When the State is the owner, it is difficult to make distinction between economic rationality and public pragmatism. Public pragmatism may demand various measures which conflict with economic rationality. And, when that happens, a conflict always takes place, and it is the electricity board which suffeRs. As for instance, in providing almost free power to agriculture and not getting a subsidy upfront to meet the cost of power distributed to agriculture. Again take the case of Karnataka. It cost the electricity board almost Rs. 2 per unit to supply power to agriculture sector, the average realisation from that sector in terms of whatever tariff exist today, is about 22 paise per unit. Even that is not collected because of agitation and so on. So, as a result, every unit of power distributed to agriculture, the loss is about Rs. 2 per unit. About Rs. 1,100 crores is the extent of loss by the agriculture consumption in the State.

Then there are problems of incentives in performance. Most of the electricity boards have large work force—about 5,000 people. In the absence of incentives and in the absence of proper goal post, it is not proper to expect them to function effectively and at a high level of performance. That is the main reason why their productivity is low. Under the

State system, it is next to impossible to increase the productivity of the cutting edge without incentives; without some accountability; and without, one can cruelly call, the hire-and-fire system. As a result, there are severe organisation and managerial deficiencies that come in the way of providing quality power to consumers.

In this background, the object of reforms and accomplishing the same has to be considered. The object of reforms is actually three-fold. The *first* is to provide efficient service to the consumers, a concept supported by all utilities. The *Second* is to ensure that power is given at a minimum price. In order to minimise the price of delivery, the supplier has to ensure efficiency in generation; minimise loss in transmission and distribution; and effective organisational performance. And finally, in order to provide new investments in this sector, not only in generation but also in transmission and distribution, it is imperative that the country creates a fiscally responsible system by which investors will feel confident of getting their returns and also some protection from risk. The Electricity Regulatory Commission (which will separate the setting of tariffs from other considerations) will eventually move towards privatising distribution networks. Orissa has adopted this model. This approach has become inevitable. And, it the country does not move towards this reform, then there is every likelihood that our system will collapse sooner than we think.

IMPACT OF ENERGY CONSERVATION IN ENERGY SECTOR DEVELOPMENT

29

K.K. Dhingra
Executive Director, Petrolium Conservation Research Association,
New Delhi

INTRODUCTION

With the growth in Indian economy, the consumption of commercial energy has also gone up. The share of commercial energy which was 40% in 1960-61 is expected to go further to 80% by the year 2004-05 due to Industrialisation, Population increase, Urbanisation and growth rate etc.

Assuming, that the growth rate in the above areas remains at the present level, it is estimated that the energy requirements would quadruplicate in next 20 yeaRs. This also assumes that no additional incremental effort is put in for energy conservation. In case concerted efforts are made by various agencies, the energy requirements would reduce considerably as it has happened in several developed countries after the oil crisis. This will also reduce the incremental investment. Hence, implementation of well thought-out policies and programmes on energy conservation is essential.

The energy intensities in several developed countries have declined sharply since the first oil shock of 70s. While, in most of the developing countries the energy intensities continue to be on the higher side. The energy consumption in KGOE per $ of GDP is one of the highest in India at 0.77 against South Korea 0.36, USA- 0.31, UK- 0.22 and Japan- 0.1. Thus, there is significant scope for energy conservation. This not only calls for continuation and reinforcement of the conservation measures already introduced by various Governmental agencies but also for adoption of new and innovative approaches. Several energy conservation studies by working groups and experts have revealed that our energy use efficiency is low compared to that achieved by the industrially advanced countries. The Inter Ministerial Working Group has estimated the energy conservation potential of 20-30% in all sectors of Indian economy.

ENERGY SECNARIO IN INDIA

In recent years, the energy sector in India has been the centre of much debate. The energy requirements in the various sectors are met through a combination of commercial energy i.e. electricity, coal, oil, gas etc. and also non-commercial energy in various forms. In India, commercial energy consumption is concentrated in the urban and semi-urban areas for Industrial, Transport, Commercial and domestic use. The non-commercial energies is widely dispersed primarily in rural areas in the form of firewood, agricultural residues, animal waste, and human and animal power.

The energy scenario in India is rather complex and needs systematic evaluation. Nearly 30 percent of the total outlay of the Government of India is earmarked for the energy sector. Inspite of such an investment in the development of Oil, Gas, Coal, and Power, the gap between energy availability and demand has not declined.

CHALLENGES AHEAD

Considering limited financial and energy resources and the substantial scope in improving energy efficiency, the following major challenges emerge :

- Efficient use leading to conservation and environmental protection.
- Strict demand management
- Reduced dependence on oil by inter-fuel substitution
- Development of alternate and renewable sources of energy etc.

Above mentioned issues are interrelated and need collective efforts to achieve the desired goals.

OIL CONSERVATION EFFORTS

In India, the focus on energy conservation in the petroleum sector was initiated with the setting up of the Petroleum Conservation Action Group (PCAG) on 6th January 1976; On August 1978, PCAG was registered as "Petroleum Conservation Research Association (PCRA)" a wing under the Ministry of Petroleum and Natural Gas with the following objectives:

- To create awareness among the people about the importance, methods and benefits of conserving petroleum products and environmental protection.

- To promote research and development efforts aimed at petroleum conservation in various sectors of the economy.
- To promote substitution of petroleum products with alternative sources of energy.
- To promote environment protection by development and deployment of fuel efficient and green technologies, equipment and devices for industrial and transport sectors of the economy; and
- To provide training and technical advisory services designed to achieve economy and efficiency in the use of petroleum products for a cleaner environment.

In order to be effective and to bring in synergy with better reach to the highly dispersed target groups in sectors like Industrial, Transport, Agriculture and Domestic, efforts were also focussed with external organisations by adopting the novel concept of Institutional networking For each sector, linkages have been established with organisations sharing similar objectives and concerns as those of PCRA. This has helped considerably in improving PCRAs effectiveness and reach. Performance of PCRA in the four major sectors of the economy is given below :

Industrial Sector

Industrial sector has registered an impressive growth rate of 12.6% during the last decade. During the initial phase of activities PCRA observed that this sector is easily amenable for time bound energy conservation measures. In India there are large number of small and medium enterprises (SME's) where there is a substantial scope for energy efficiency improvements. Most of such Industries are in clusters. Catering to these clusters can bring about substantial savings and Energy audit and accounting can be gainfully employed. Better housekeeping, proper maintenance, efficient management of steam, heat and electrical loads are some of the key areas for efficiency improvement. Waste heat recovery systems, cogeneration and utilization of alternative resources of energy are other important measures which also enhance energy conservation. Further, Technology upgradation, modernisation and introduction of control instrumentation realize the full energy conservation potential. In certain basic industries, process modifications or adoption of new technologies are also required. Thus, to tap the full potential for energy conservation and to protect the environment in the SME's calls for close co-ordinated work between industrialists, technologists and consultants.

As on March 1997, PCRA had carried out fuel oil diagnostic studies and energy audits in 3865 industries. These units have a total annual fuel oil consumption of about 7 million kls. These studies have identified a saving potential of 778000 Kls per annum which is

10.95% of annual fuel oil consumption. After submission of energy reports, PCRA closely follows up for implementation of the recommendations so as to bring about improvements in energy utilisation. Such follow up studies convert potential identified in energy use studies to real savings. As a result, industries serviced by PCRA have realised a saving of about 6,86,000 Kls on a recurring basis through implementation of PCRA recommended measures. PCRAs success rate for converting Oil saving potential to saving realised varies from 82% in process industries to 98% in Textile industries depending upon the type of Industries listed as follows.

Table 29.1 : Results of Studies Conducted by PCRA

(All quantities in 100000 kl)

Type of Ind.	Units Surv.	Annual Consu.	Saving Potnt. Ident.	Pol. Vs % Cons.	Saving Real. (Qty)	Saving Real.%
Engg.	1577	20.09	2.265	11.27	2.01	88.74
Process	1570	29.36	2.881	9.81	2.370	82.26
Textile	371	7.33	0.75	10.23	0.74	98.6
Glass	174	4.15	0.374	9.01	0.344	91.97
Misc.	173	12.276	1.51	12.3	1.4	92.71
TOTAL	**3865**	**71.046**	**7.78**	**10.95**	**6.864**	**88.17**

From the analysis of the energy use studies conducted by PCRA, it is observed that 54 percent of the total saving potential can be realised with nil or marginal investment simply by adopting better housekeeping, operational and maintenance practices. The remaining 46 percent of the saving potential needs marginal capital investment. High cost energy efficiency improvement projects involving change in technology are excluded in energy use studies. A summary of oil conservation measures identified by PCRA during its energy use studies and estimated percentage saving potential in Indian Industries is given below:

In addition to the Energy use studies mentioned above, PCRA has also instituted promotional schemes for promoting efficient utilisation of Petroleum products :

Promotional Schemes/Demonstration Programmes for Oil Conservation

To promote oil conservation, PCRA has successfully launched the following major schemes and programmes.

Conservation Measure and Percentage oil Saving Potential

Conservation Measure	*% Saving Potential*	
• Storage, handling and preparation of fuel	5.5	54
• Draft Control in boilers and furnaces	25.8	
• Insulation	8.4	
• Condensate/Flash steam recovery/ plugging of steam leaks etc.	14.3	
• Waste heat recovery	19.0	46
• Other measures (boiler replacement, process modification etc.	27.0	
	100.0	

- *Empanelment of Energy Auditors*—Based on the expertise gained over the years, PCRA assesses, evaluates and develops potential Energy AuditoRs. A committee of renowned energy audit experts has been constituted to promote, develop and on the basis of set criteria empanel energy auditors throughout the country. Around 43 energy auditors have been empanelled under the scheme.

- *Energy Audit Subsidy*—PCRA offers subsidy upto 50% of the cost towards conducting energy audit at the party's industrial premises limited to a maximum amount of Rs. 50,000 per industrial unit through PCRA empanelled energy auditors.

- *Inefficient boiler replacement scheme* — Soft loans @ 8 per cent interest per annum is given to the Industries by the Oil companies for boiler replacement of inefficient oil fired boilers (having an efficiency of less than 70%) with efficient oil/coal fired boileRs. Under the scheme around 360 inefficient boilers have been replaced with efficient ones.

- *Development and commercialization of low excess air film burner*— Over 4000 such burners are in use by different industrial units resulting in 10-15 percent fuel saving.

- *State level actions groups*—PCRA assist in setting up of State level action groups co-ordinated by State level co-ordinators involving government authorities, oils companies, industries etc wherein policy decisions on conservation activities are discussed to evolve and implement action plans for conservation.

- *Energy Auditors award*—In order to promote energy auditing and to motivate, recognize and give thrust to the conservation activities of Energy auditors, PCRA has evolved a scheme for the empanelled energy auditors under which an award is conferred on the best two energy auditor in the form of silver plaque trophies with appropriate citations and a cash award of Rs. 20,000 and Rs. 15,000.

Transport Sector

In view of substantial scope in the transport sector, the emphasis is given parallely on several fronts i.e. on the training of drivers and driver instructors/trainers in acquiring and applying improved driving skills, establishing model depots and Model garages, development of fuel efficient engines, re-refining and recycling of used lubricating oil, promoting high performance oils, instituting diesel conservation awards, setting up of State level Action Groups and making them result-oriented, development and promotion of use of fuel-efficient technologies, inter-fuel substitution etc.

In addition to the activities mentioned above, PCRA has also instituted promotional schemes for promoting efficient utilisation of Petroleum products :

Promotional Schemes/Demonstration Programmes for Oil Conservation

- Soft loan for replacement of inefficient engines with fuel efficient engines. The scheme encourages use of modern fuel efficient diesel engines (145-165 GMS/BHP/HR) against the conventional diesel engines (180-190 GMS/BHP/HR). Under the scheme, around 944 engines have been replaced. Since the manufacture of the inefficient engines has been stopped and the desired objective is achieved therefore the scheme has been discontinued.

- Setting up of model depots—PCRA assesses the maintenance practices of the Garage/workshops based on which it gives recommendation to upgrade the services rendered by the depot to a Model depot. Special stress is laid to improve the maintenance practices and to introduce better management control systems. This is a five phased programme comprising of selection of depot, diagnostic studies, implementation, driver training, follow-up and project evaluation report presentation to demonstrate benefits achievable through adoption of better maintenance practices and tuning of vehicles.

  Setting up of model garages to upgrade operational and maintenance facilities to demonstrate benefits achievable through adoption of better maintenance practices and tuning of vehicles. Under the scheme, several instruments/equipments have been identified which effectively help in achieving conservation of fuel. Financial assistance is provided upto Rs. 10 lakhs or 50% of the cost of the instrument/ equipment whichever is lower at a nominal interest of 8% per annum.

- State level action groups—PCRA assist in setting up of State level action groups co-ordinated by State Level co-ordinators involving government authorities, oils companies, industries etc wherein policy decisions on conservation activities are discussed to evolve and implement action plans for conservation.

The above efforts have led to improvement in the average KMPL in the organised sector from 4.01 to 4.43 within a span of 15 years.

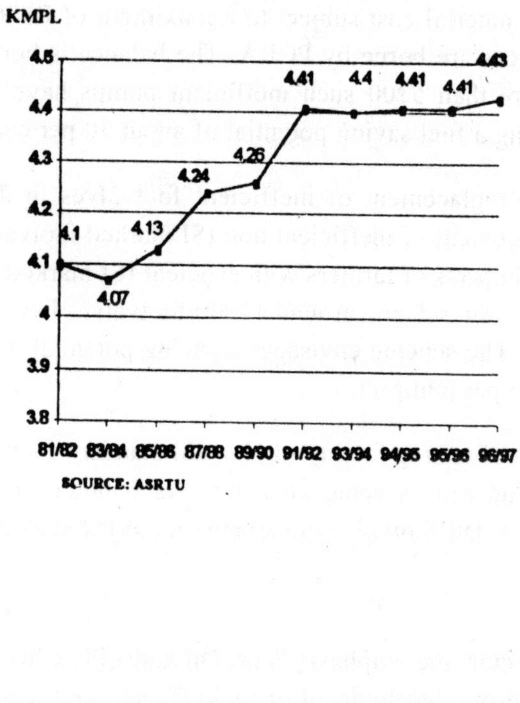

SOURCE: ASRTU

Agriculture Sector

In the agricultural sector, the primary focus is to promote awareness amongst the farmers most of whom are not educated by setting up of diesel lift irrigation pump demonstration centres. This includes practical demonstration on the benefits of fuel efficient ISI marked pumping systems over the inefficient non ISI marked pumping systems, partial or complete rectification of diesel-powered fuel-inefficient irrigation pumpsets, replacement of fuel-inefficient foot valves by those bearing ISI mark. All-out efforts are made to reach the farmers through exhibitions, kisan melas, workshops, clinics etc apart from reaching and educating farmers through State van publicity, on better operation and maintenance of tractors, irrigation pumping sets and other agricultural implements.

In addition to the activities mentioned above, PCRA has also instituted promotional schemes for promoting efficient utilisation of Petroleum products :

Promotional Schemes/Demonstration Programs for Oil Conservation

- Subsidy for rectification of diesel irrigation pumping system—To demonstrate to farmers the savings accruable through rectification of defects in diesel operated irrigation pumpsets. The scheme ennvisages replacement of fuel inefficient footvalves, suction and delivery pipes with that of fuel efficient ISI marked system leading to a saving of 30% in terms of fuel consumption per LIP. Under the scheme 50% of the material cost subject to a maximum of Rs. 600 per pumpset and fuel service charges are borne by PCRA. The balance is borne by the beneficiary. As on date more than 5200 such inefficient pumps have been rectified by PCRA, demonstrating a fuel saving potential of about 30 per cent.

- Subsidy for replacement of inefficient footvalves in diesel pumpsets—Scheme covers replacement of inefficient non ISI marked footvalves of diesel operated lift irrigation pumpsets of farmers with efficient ISI marked footvalves (friction factor < 0.8). Under the scheme around 12000 footvalves have been replaced throughout the country. The scheme envisages a saving potential of 8-10% in terms of diesel consumption per pumpset.

- Upgradation of testing facility of Non ISI standard footvalve manufacturers to ISI standards. Under this scheme a loan upto Rs. 65000 is granted to upgrade the testing facility of non ISI footvalve manufacturers to ISI standards.

Domestic Sector

In the domestic sector, the emphasis is on Oil and LPG conservation by better cooking habits. PCRA sponsors development of fuel-efficient kerosene wick stoves, LPG stoves and domestic lighting appliances using kerosene apart from promoting their use. To promote substitution and direct savings of kerosene and LPG, PCRA also encourages the use of alternative sources of energy such as solar, bio-gas, etc.

- Development and promotion of fuel efficient LPG/kerosene cooking stoves, etc.—These fuel efficient stoves result in 20-25 per cent fuel saving compared to ordinary cooking stoves available in the market. About 80 lakhs efficient kerosene wick stoves and 5 lakhs LPG stoves are in use in the country.

- Better driving habits—PCRA organises workshops/clinics for the youth, housewives etc. on tips for better driving of 2/3/4/wheelers used in the domestic transport.

Research and Development

Realising that any technological breakthrough and its successful commercialisation will have considerable impact on the economy, PCRA promotes and encourages research and development efforts by networking with several R&D establishments, CSIR laboratories, technical institutions, etc. for:

- Development and promotion of fuel efficient appliances and processes;
- Evaluation of additives/devices/appliances for fuel economy;
- Development of instruments for energy efficiency monitoring;
- Energy efficiency Optimization by retrofitting and modifications; and
- Substitution of petroleum products with alternatives fuels.

Multi-Media Awareness and Educational Campaigns

PCRA aims at making oil conservation a national movement. To take the message to the people, it explores the potential of and uses all possible and effective media for mass communication. This includes electronic media e.g. T.V., Radio, electronic displays; press at the national and State level; printed literature for specific target groups; outdoor publicity through Hoarding, Bus panels, Kiosks, Balloons, Banners, translides, TV spots, Radio talks, Jingles etc. The focus of all messages is easy to implement and practical conservation tips for the Industrial, Transport, Agriculture and Domestic sectors. For effective communication to the target groups in semi urban and rural areas, messages are made in regional languages.

Oil Conservation Fortnight

The oil conservation week was launched in January 1991. Inspired by the overwhelming response from the government agencies, non-governmental organisations, and voluntary organisations among others, it was decided to extend the conservation activities to a fortnight. Under the aegis of the Ministry of Petroleum and Natural Gas (MOP&NG), the entire oil sector joined PCRA in making the 8th Oil conservation fortnight from 1-14

February 1998 a great success. The theme of OCF '98 was "OIL CONSERVATION FOR CLEAN ENVIRONMENT". As a curtain raiser to the OCF '98 in New Delhi, a rally of over 5000 school children was held at India Gate, New Delhi. It was flagged by the then Lieutenant Governor of Delhi, Shri Tejendar Khanna for spreading the message of Oil conservation particularly in the rural and semi-urban areas.

ACTIVITY NO

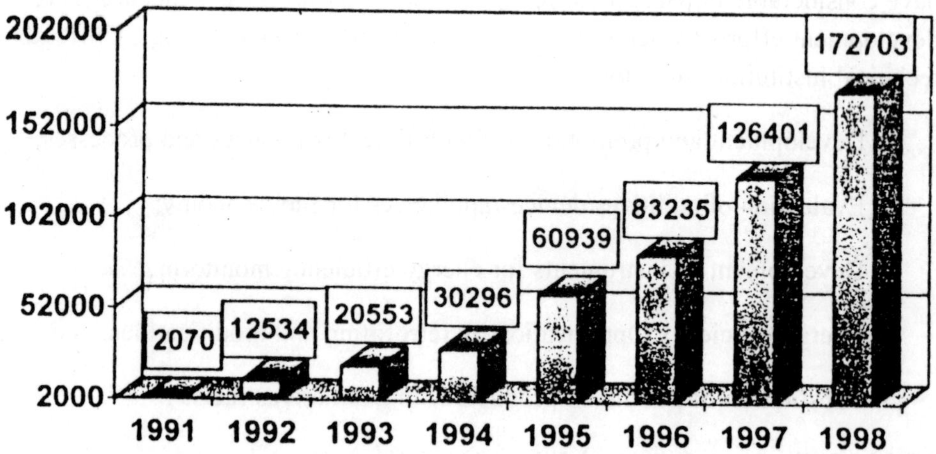

Impact and Achievements on Oil Conservation

Co-ordinated efforts by the oil industry and large number of networking institutions have resulted in substantial petroleum products savings as shown below:

Conservation Targets for Ninth Five Year Plan

The Ninth Five Year Plan proposes curtailing the specific energy consumption level by 15% with respect to GDP in all the sectors of the economy. The savings estimated during the Ninth Five Year Plan period, in terms of reduced consumption of petroleum products by around 2.7 million tons is valued at about 2720 crores. The following major strategies are being adopted.

1. Policies to make energy conservation attractive by providing incentives to users.
2. To bring attitudinal changes amongst energy users through appropriate legislative measures and take steps to discipline those who fail to adhere to these changes.

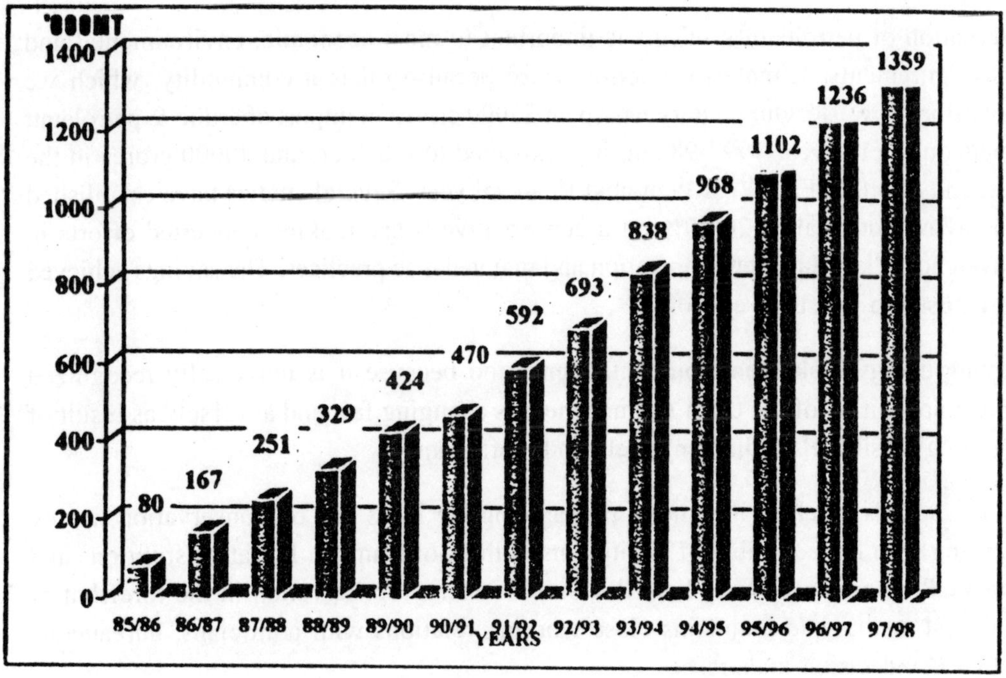

3. Mandatory energy labeling of equipment, appliances, techologies and processes. Making mandatory the use of energy efficient equipment's in industries and display the energy efficient parameters of the equipment as specified and make it a part of the Industrial License. Phasing out inefficient equipment by prohibiting the manufacture and sale of the same. To make energy audit mandatory. Establishing energy efficiency standards for equipments, appliances and process in line with international standards and enforcing.

4. Establishment of petroleum fuel consumption norms for different industries. All projects for Government approval to be acompanied by an energy plan and to be cleared only when conservation aspects are fully safeguarded.

5. Effective demand management.

6. To view energy conservation as a source of energy, and use it to bridge the gap between demand and supply.

7. Investment in small and medium scale industries for technology upgradation.

CONCLUSTION

Conservation of petroleum products is important to meet economic, environmental and social requirements. It makes economic sense because oil is a commodity, which we cannot afford to waste. Our country has spent 33000 crores in import of crude & petroleum products during the year 1997-98 and it is expected to touch around 40000 crores in the project end of Ninth Five Year Plan next financial year. Several studies have established an oil saving potential of 20-30% on a conservative basis, making concerted efforts in improved house keeping, better operation and maintenance practices. The savings achieved could be used to boost the economy.

Saving energy makes environmental sense too because it is universally recognized that the composition of the earth's atmosphere is changing fast and adversely as result of our burning fossil fuels indiscriminately and inefficiently.

PCRA spends much of its time in propagating the ideas and oil conservation tips by networking with large number of institutions in the Government, Private institutions and NGOs. PCRA has successfully kept the issue of energy conservation in the forefront of the national media, and maintains close working relations with politicians, bureaucrats and other key decision-makers.

FINANCING OF ENERGY PROJECTS INVOLVING INNOVATION, TECHNOLOGY TRANSFER AND ENVIRONMENT

30

O.P. Rao
Deputy Adviser,
Council of Scientific and Industrial Research,
New Delhi

INTRODUCTION

India is endowed with primary energy resources in the form of water, fossil fuels (coal, lignite, oil and natural gas) and nuclear fuel as well as non-conventional sources of energy such as wind, solar, biomass, and geothermal etc. These energy sources have been developed to different degrees and lot of potential remains to be exploited. All the energy resources have to be exploited to the extent possible to meet the ever increasing energy demand of the country which is growing at 6.3% per annum.

There are basically 3 types of strategies for setting up energy projects for utilisation of energy resources.

1. Application of existing technologies
2. Upgradation of existing technologies
3. Introduction of new technologies

All the three strategies will be required for efficient and effective development of energy sector.

In the first strategy the technology is well established and commercially proven, hence the issues pertaining to innovation and technology transfer are not relevant in financing of the project. Examples in this category are: large scale hydroelectric projects and pulverised coalfired thermal power plants. Financing of such projects is mostly from the Government budgetary resources as a part of infrastructure development through implementation of annual or Five Year Plans. Power sector is also opened up for private sector participation/foreign investment. There are set procedures and guidelines for financing private power projects such as signing of power purchase agreement (PPA) and fuel supply agreement (FSA), guarantee and counter guarantee, construction contract with equipment supplier and EPC contractor, environmental clearance and willingness of bankers/financial institutions for lending the money etc.

The second category i.e, upgradation of existing technologies mostly consist of modernisation and renovation, repowering and retrofitting. Examples in this category are again hydroelectric projects and coal fired thermal power plants. Most of these technologies are proven on commercial scale to be applied for the operating commercial plants. If the returns on investment are attractive, financing of the projects will be possible. Upgrading can be done by the Government or by the private sector.

The third category pertains to new technologies involving innovation, technology transfer and environmental issues. Demonstration or Technology Proving Plants come under this category. Demonstration is an activity in the innovative chain and an important step in technology transfer. Financing these projects by regular lenders poses problems due to very high risk involved in the commercial viability. Various mechanisms and issues involved in financing demonstration projects are discussed in this chapter.

PROBLEMS IN FINANCING INNOVATIVE PROJECTS

* High risk in success/commercial viability of the technology.
* Capacity of the demonstration plant is generally below the scale of economy, hence operation of the plant may not be technoeconomically viable.
* The plant may not generate enough cash flows required to provide satisfactory returns to the investor.
* Decision making on investment specially in the Government sector is difficult and time consuming due to involvement of too many agencies/Ministries and for fear of criticism later if the technology fails. Due to delays serious partners withdraw from the project.
* The plant is first of its kind, hence experience and expertise on design, specifications, fabrication, erection, commissioning and operation is not available. This can bring in some uncertainty in the cost and time of completion.
* Inconsistency in policy due to change in the decision making officials or political changes.

* Issues related to confidentiality, credit sharing, intellectual property right and ownership of technology, if not settled amicably, can create problems.

MECHANISMS FOR FINANCING OF INNOVATIVE PROJECTS

In India R&D is mostly supported by the Government budget which is on the decline, however investments in the private sector are going up. R&D is expensive and needs multidisciplinary expertise and infrastructural facilities, which may not be possible to have all of them in one organisation. This calls for joining together of various organisations to pool their resources for working on the innovative projects. An innovative project consists of R&D work at several stages starting from laboratory scale, pilot scale, demonstration project and ends with setting up of commercial scale plant. In this innovative chain demonstration project is very significant since it proves the commercial viability of a technology developed on a laboratory/pilot scale. Demonstration project also requires more funds than laboratory/pilot scale plant.

Government-Industry Partnership

Joint financing of an innovative project by Government and industry is illustrated by taking the example of the "Clean Coal Technology Demonstration Programme" (CCT Programme) being implemented in USA. The CCT programme is composed of 40 demonstration projects that will bring to the market place a range of advanced cost effective and environmentally benign technologies for use of coal. It is a unique partnership between the United States Department of Energy, (USDOE) and Industry. The industrial partner contributes atleast 50% of the total cost of project. The programme commenced in 1985 and it has successfully achieved many milestones so far. Need for taking up the programme, its objectives and factors for success of the partnership are discussed below.

Need for taking up CCT programme was due to the following reasons

* Coal is to continue as the major contributor to USA's energy supply mix with reserves of 297 billion tonnes which is 28% of world coal reserves, 58% electricity and 25% primary energy coming from coal.

* Planners of the country have recognised that continued viability of coal as a source of energy was dependent on demonstration and commercial application of new generation of advanced technologies characterised by enhanced technical, economic and environmental performance.

* Improving the perception of coal as a low cost and environmentally acceptable fuel is necessary to promote coal in place of oil and gas.

* To fulfil the clean air act amendments of 1990 (CAAA) and to enhance environment security.

* A global market potential of $ 200 billion exists by the year 2015 for US technology suppliers, equipment manufacturers, and engineers etc. and can also generate 100,000 jobs in the domestic power equipment industry.

* To export coal of the order of 120 million tonnes by the year 2015.

Objectives/goals of the programme are.

* Creating a cleaner environment with coal to fulfil the 1990 clean air act amendments.

* Fuelling an expanding economy (US needs 150,000-200,000 MW of new capacity by the year 2010).

* Creating productive jobs for Americans.

* Making US secured in energy.

* Making US more competitive in the world market for clean energy from coal.

* Creating a new Government-Industry partnership

Main reasons for success of Government-Industry partnership were as follows:

* The partnership is based on cost sharing principle.

* Roles of the Government and industry are clearly defined. Industry is responsible for technical management while the Government aggressively monitors the projects.

* A strong and stable financial commitment exists for the life of the projects. Government releases funds on a time phased basis.

* Role of the Government is that of a risk sharing partner.

* Industry is fully committed to the programme by investing two dollars for every dollar spent by the Government. Total cost of CCT programme is 5.7 billion dollars out of which Government (DoE) and industry share is 1.9 and 3.8 billion dollars respectively.

* The technical agenda (i.e. selection of projects) was determined by the industry not the Government as per the policy and guidelines established by the Government.

* Demonstrations are conducted at commercial scale in actual user environment and under conditions typical of commercial operations which allow the potential of technologies to be evaluated in their intended commercial applications.

* Industry retains real and intellectual property rights, which encourages industry to come forward for investment and promotes commercialisation.

* Industry must make a commitment to commercialise the technology and the technology must be made available to the potential domestic users on reasonable commercial terms.

* Repayment of the Government share by industry is required only upon successful commercialisation of technology. Different sets of rules are formed for recoupment: from revenue streams derived from the operation of the demonstration plant beyond the demonstration phase and the commercial sale, lease, manufacturing, licensing or use of the demonstrated technology and upto 2% of gross equipment sales and 3-5% of royalties.

* Information generated from the demonstration projects is widely disseminated to users, equipment manufacturers, policy makers and the public through conferences, newsletters, updates and internet.

Cooperative Research

Unique example of cooperative research is the Australian cooperative Research Centres programme under which the "Cooperative Research Centre for New Technologies for Power Generation from Low-Rank Coal" was established in the year 1993 to conduct research into aspects of those new generation technologies and processes that have the best prospects of overcoming the principal challenges (cost and environmental impact) facing the future use of low rank coal as a competitive energy source for electricity generation. New generation technologies are the key to the objectives of reducing both costs and emissions to the environment.

The formal agreement between the nine participating organisations and the Commonwealth establishes the Centre as an unincorporated joint venture. The Agreement commits some $ 33 million over seven years to the activities of the Centre with the Commonwealth contributing $ 13 M, industry contributing $ 7 million in cash and the balance being provided by participants in the form of in-kind contributions. This provides a significant resource base on which to build the Centre's programmes. It also provides necessary support for the export of technologies.

The centre is functioning as a joint venture. The research directions of the centre are determined by the needs of the industry and the requirements for commercial application. Several potential end users are participants in the centre which ensures commercialisation of technologies.

Other Schemes of Financing

It is possible to get funds for innovative projects from international and national agencies. The international agencies are GEF, USAID, UNDP etc. With reference to GEF, the energy projects specially coal related come under the focal area of "Climate Change" " and under the operational programmes of "Removing barriers to energy conservation and energy efficiency" and "Reducing the long term costs of low greenhouse gas emitting energy technologies". GEF portfolio in India is diverse and varied and as on January 1998, a total of US $ 142 million has been programmed for India. However GEF provides only limited funds towards incremental costs which amount to less than capital cost of the project.

Financial institutions in the country like Industrial Development Bank of India (IDBI) provides venture capital for innovative projects and during the year 1997-98 IDBI sanctioned Rs. 49.4 crore under venture capital fund. The National Research and Development Corporation (NRDC) also provides venture capital. NRDC finances Technology Development Projects in collaboration with industry/R&D institute for setting up pilot/semi commercial/demonstration plants. Equity participation is also undertaken by NRDC. The corporation also undertakes joint development projects with industry by sharing the cost upto 50% and the rest is put by industry. If the project is successful industry pays back the corporation's share of expenditure, however if the project is unsuccessful the corporation writes off its share of expenditure completely, funds available under these programmes are of the order of 10-20 crores in one year.

The Department of Scientific and Industrial Research has a programme called "Programme Aimed at Technological Self Reliance (PATSER) which supports technology development and demonstration projects, outlay for Ninth Plan proposed is about Rs. 60 crore. The Technology Development Board under the Ministry of Science and Technology provides soft loan at 6% interest for commercial application of indigenous technology or imported technology. In addition to loan there is provision to give financial assistance in the form of grant and equity. The repayment of loan will commence after one year of successful completion of project. The availability of funds for 1997-98 are likely to be Rs. 70 crore.

Technology Information Forecasting and Assessment Council (TIFAC) under its activity "Home Grown Technology" is promoting commercialisation of indigenous technologies.

Financial assistance is provided for development of technologies from lab scale to commercial scale. Risks and uncertaintics are shared by TIFAC. Repayment is spread over a period of 5 years after the gestation period on easy terms as soft loan. A provision upto Rs. 20 crore per annum is likely to be available under this scheme.

Another financing scheme for technology development jointly by industry and national R&D institutions is called SPREAD (Sponsored Research and Development Programme) promoted by the ICICI. Over 100 companies have been provided assistance so far under this programme including energy sector. Amount sanctioned in these programmes exceed Rs. 50 crore per annum.

There are a number of mission mode programmes being implemented jointly by Government and industry for example sugar technologies, advanced composite, flyash, leather. In these projects finance is shared between Government and industry. Often the equipment are installed at the industry at the technology development phase itself thus cutting down time for commercialisation.

All the five IITs and IISc Bangalore have come together to take up, in partnership with industry, Technology Development Mission (TDM) projects in seven generic areas. Each Mission project is to have atleast one industry partner, who will contribute a minimum of 25% of the total cost. In return, the industry partner will have the first right to commercially exploit the outcome of the project, which are to have identifiable deliverables in the form of hardware, software, process or product know-how etc.

Government is taking measures to orient and encourage the national laboratories to work with industries and users as well as to seek funds from them. In this direction, Government is providing a matching grant to the national laboratories for every incremental commercial earning from industries. This will benefit both lab and industry. Lab gets additional funds and industry could get 2 rupees worth of R&D work for every rupee invested by them.

Government has also introduced a few important direct benefits to industries in the form of tax concessions and duty exemptions which are means of financing technology developments. There are tax holidays to the companies engaged in R&D. Customs duty exemption is available on specified goods imported for use in R&D projects. Various excise duty exemptions are also available.

FINANCING OF IGCC DEMONSTRATION PLANT IN INDIA—A CASE STUDY

IGCC i.e. Integrated Gasification Combine Cycle is a clean coal technology for generation of power. When compared to the conventional pulverised coalfired thermal power generation, IGCC is more efficient and emits negligible pollution even while using high ash coals. A mission programme has been taken up for setting up an IGCC demonstration plant in the country to prove the technology with Indian coals. It is an energy development project involving innovation, technology transfer and environmental issues.

The objectives of the project are as follows

1. To demonstrate operation of IGCC and evaluate the technoeconomics.
2. To demonstrate the higher overall thermal efficiency of the IGCC plant compared to conventional thermal power plant.
3. To establish the improvement if any in the average plant availability compared to that of conventional coal based thermal power plant.
4. To demonstrate reduction in atmospheric pollution of IGCC plant compared to those from an existing/nearby best operating conventional power plant in India.
5. To draw a plan for commercial application.

The financing plan proposed for the demonstration plant is as follows:

The IGCC demonstration plant of 100 MW capacity is proposed to be setup in an existing thermal power plant (TPP) so as to utilise its available facilities in order to reduce the cost of the demonstration plant. Capital cost of the demonstration plant is estimated as Rs. 580 crore based on the cost data for IGCC plant available with several organisations in the country. This is a very rough estimate since data on IGCC plant similar to Indian plant is not available. Because of higher uncertainty contingencies have been taken as 10%. Similarly cost of engineering and establishment has also been taken as 10%. Since this is a R&D project for demonstrating a technology, import duties are not considered.

TPP is willing to contribute 50-70% of the cost of the IGCC plant. The Technology Development Board (TDB) under the Ministry of Science & Technology is expected to provide upto Rs.100 crore as a soft loan @6% interest. Since the project is a clean coal technology, grant is expected from international agencies like GEF, and Green Aid of Japan and from the Government of India e.g. Ministries of Power, Coal and Environment. This has to be a grant to reduce fixed cost of the project and cost of generation. Two financing plans have been drawn based on investment by TPP at 50% and 70% of the cost with debt equity ratio of 70:30. Incidentally cost of a pulverised coal fired plant is roughly equal to 70% of the cost of a IGCC plant. Thus the financing of the project consists grant, soft loan, debt and equity.

The following two financing plans have been suggested :

(Rs. crore)

	Fi n. Plan-I	*Fin. Plan-II*
Capital cost	580	580
TPP contribution	290 (50%)	406 (70%)
TDB loan	100	100
Grant	190	74
Cost of generation, Rs. /Kwh	2.4	2.9

In the Financing Plan-II cost of generation is high because the amount of grant is reduced. The cost of generation of an equivalent capacity PC plant is comparable with the value in Fin. Plan-I. In the Fin. Plan-II subsidy is required on cost of generation to make it comparable with a PC plant.

An IGCC plant consists of a coal gasification island and a power island. The coal gasification island comprises of a coal gasifier, and gas cleaning system which is approximately 30-40% of the total cost of the IGCC plant. Coal gasification process for power generation is yet to be proven with high ash coals, so this is the component of risk involved in IGCC demonstration plant. The power island comprises of a heat recovery steam generator, steam turbine and a gas turbine etc. which are proven equipment in power generation.

In case of failure of IGCC demonstration plant, most of the equipment in power island costing about 60-70% of total plant cost can be utilised in other power plants. Thus if the Government is prepared to finance the risk component in IGCC demonstration plant to the extent of 40-50% as grant, it will be possible to mobilise the remaining resources from other parties interested in the commercialisation of technology in the country and the project could also be economically viable.

Another possibility regarding financing of the project is contribution in kind from the equipment manufacturer. In this mechanism, gas turbine and steam turbine can be supplied by the respective manufacturers on credit basis to be recouped after the plant is completed and revenues are generated from the sale, and licensing of technology or from sale of power produced from the plant. Coal for demonstration purposes can be supplied free of cost by the Ministry of Power.

CONCLUSIONS

Though research is carried out in the national laboratories funded by the Government, ultimate user is the industry. Involvement of industry with research work, preferably, right from the stage of selection of project will help in the commercialisation of technology and to a certain extent lessen risk in the success of the project. Financial participation by the industry makes him a serious partner committed to the project and reduces the Government funding. Participation of industry encourages and motivates research personnel working on the project.

Risk and uncertainty of an innovative technology can be considerably reduced by demonstration prior to commercial application. Innovation and demonstration projects in the energy sector are expensive requiring large funds, often beyond the financial capability of a single organisation. Government funding towards R&D is decreasing. In these circumstances multi agency funding is a proper financing mechanism. Government gives customs and excise duty exemptions and tax concessions for the promotion of R&D projects. In addition to these benefits incentives and subsidies may also be considered. Once the commercial viability of a new technology is demonstrated, the market forces could step in to provide finance for IGCC commercial projects.

Financing of commercially established energy projects is relatively simple, but when the issues pertaining to innovation, technology transfer and environment are involved, financing becomes difficult and complicated. A single model is not valid for financing all the demonstration projects. However, Government financing towards risk and atleast 50% of the cost of the project by industry are required for setting up of demonstration projects.

The Government-Industry partnership for clean coal technology programme followed in USA appears to be the most suitable financing mechanism for demonstration projects.

REFERENCES

1. Approach Paper on Setting up an Integrated Gasification Combined Cycle Demonstration Plant.
2. Cooperative Research Centre, New Technologies For Power Generation From Low Rank Coal. Australia. Annual Reports 1993-94 and 1994-95.
3. Department of Science and Technology. Annual Report 1996-97.
4. Department of Scientific and Industrial Research, Ninth Five Year Plan (1997-2002) and Annual Plan 1997-98, New Delhi.
5. Industrial Development Bank of India. Annual Report 1997-98.
6. National Research Development Corporation. Annual Report 1995-96.
7. Planning Commission. The Standing Committee on Industry-Institution. Interaction—An Approach Paper. Promotion of Interaction Between University-Industry-National R&D Labs-National Academies.
8. Rajan Y.S. Technology Financing.
9. US Department of Energy. Clean Coal Technology Demonstration Program. Update 1996-97.

SUSTAINABILITY IN DEVELOPMENT OF ELECTRICITY SECTOR IN INDIA : A CASE STUDY

31

A.B. Kottewar, FIE[1]
P. R. Kulkarni, MIE[2]
[1]Superintending Engineer (Elect) in PWD, Govt. of Maharashtra, Member IMC, WEC
[2]Executive Engineer EHV Construction Division, MSEB, Maharashtra

India is a country of rich heritage, varied custom with a heavy cultural and social bond. We can see the impact of old culture on the life of every citizen in India. World community may define India as a developing or backward country but in its own, the level of dedication in society is so high that even extreme poverty in isolated region could not deviate them from the designated path shown by ancient culture.

The sustainable development should meet the needs of the present generation without sacrificing the needs and aspirations of the future generations without endangering the ecology and environment. In this reference, the principles of sustainable development need to be applied. For the backward and developing countries, food and shelter remains the first priority in the planning of the sustainable development as far as basic needs are concerned. The development becomes unsustainable because of the collapse of a "co-evolutionary development process". This entails the elements of knowledge, values, social organisations, technology and resource systems, holistic view and multi-disciplinary approach. But, due to cultural imposition, value systems have been collapsing, knowledge has been reduced to western understanding and the social organisation and technologies have become increasingly the same around the world.

For all-round development with the optimum speed, we need the support of Electricity. The input of the electrical energy is required for modern living, industry, agriculture, transport and all-round management of development process including water management.

Considering the Indian scenario, population growth has become the main hurdle in planning the development of people. The definition of sustainability is associated with need. Need is again dependent on the resources, available in the country. The productivity is again a complex phenomena. Work culture, knowledge, intention to develop, resources, management planning, international support for technology is vital for growth. Due to shortcomings and shortages on many of the fronts connected with the sustainable development, we have to plan for the use and creation of resources. If we consider the geography of well being i.e. sustainable development, we have to consider the definition of welfare. Food and shelter are the main ingredients of the welfare which could be done in four categories.

1. Distitution, 2. Insufficiency, 3. Sufficiency, and 4. Affluence.

To provide for the sufficiency to the people in respect of the welfare, the economy has to be in tune with global situation. The speed of economic reforms is too much. To match with this, the economic growth has to be equally fast. The Electrical Energy is one of the inputs for the speedy acquirement of sustainable development.

The normal course of economic development is likely to spread the energy use patterns currently prevalent in advance economies of the world. 50% of the world's population resides in underdeveloped Asia. New demands for the energy can be satisfied from the available resources with known technologies. Thus, the real cost of energy need not escalate significantly over the projected period upto 2010. Economic growth is the main factor driving growth in Energy Demand. Between 1970 and 1993, world's Gross Domestic Product (GDP) rose from 12 trillion to 23 trillion. By 2015, world's GDP is expected to almost double again, rising to 45 trillion. The Electricity consumption is also rising with similar trend as per GDP as per the Figures in ANNEXURE II. This consolidates our projections for Electricity consumption in 2010 and 2020.

The Geographical composition of the Energy demand will change substantially over the projected period. About 66% growth is expected in newly industrialising economies of Asia.

The sustainable growth projection in India for Electricity sector is shown in ANNEXURE-I. The state of well-being and inputs are shown on graphs which relate the input to the gross satisfaction level in India. We have already said that the sustainability in the level of well-being has to be different in different countries. India, because of its place in culture/heritage has to be seen with different angle. The use of 1000 KWH per capita per year is the minimum level for forcing ahead to the developing economy. To achieve this, we could need another 80,000 M.W. capacity addition. If we decide to *Commission* this capacity by 2009. The finances for only thermal generation would be around 80,000 × 4 = Rs. 3,20,000 Crores. It is for the planners to attract the finances from developed economies to invest this amount on the acceptable or sustainable terms. Electricity is not the general commodity available for marketing. It is to be produced as per the need and to reach to consumer's door. The production, though is an important stop, would need to strengthen the infrastructure in coal mining. Transport to stations, and efficient system to handle the bulk and small power through strong network of Transmission and Distribution lines is necessary so as to give quality supply, to consumers. Today's system of bulk power transmission and small power distribution is not at all sustainable. It can not handle the power at today's per capita rate i.e. 400 KWH/capita. The voltages are improper, the frequency is nonstandard with wide variations and the reliability is poor. In such situation we propose the design of Bulk Power Transfer Network considering the 2000 KWH/Capita to 4000/KWH per Capita. A long term plan to handle such Network dependent on available land/population need to be studied and brought in practice. The total infrastructural development needs another Rs. 3,20,000 Crores and escalated amount of Rs. 4,30,000 crores. As such, India needs Rs. 8,60,000 crores to reach to 1000 KWH/capita level which is just above the insufficiency level. It is for the planners to look into this problem and to achieve the targeted economic growth in India.

CONCLUSION

At present there is over emphasis on the projection of demand by percentage and to plan for the generation capacity. India has the only resources in Thermal side and it will need One Billion ton of good coal per year upto 2010 to cope with the 1000 KWH/Capita level. Similarly the Transmission and Distribution sector which is overlooked today needs to be coupled with generation. Our aim is to use the power for the economic growth and achieve higher GDP. If we think in isolation for only one sector, it would be a suicidal mission for the Indian Economy. The world businessmen are coming forward to put the generating plants in developing countries with the intention of Brisk Trading in Electricity sector. This is accompanied with the unwarranted terms and conditions such as deemed generation. If the world at large wants to develop their marketing activities in developing economies, the developed countries and U.N.O. should come forward with the entire plan for Generation, Transmission and Distribution as a package deal. This will only boost the economy of undeveloped countries, in the real term which is need of the hour.

ANNEXURE 1

Sustainable-Levels Total-Consumptions Per Capita-per year- in KWH (Projected)

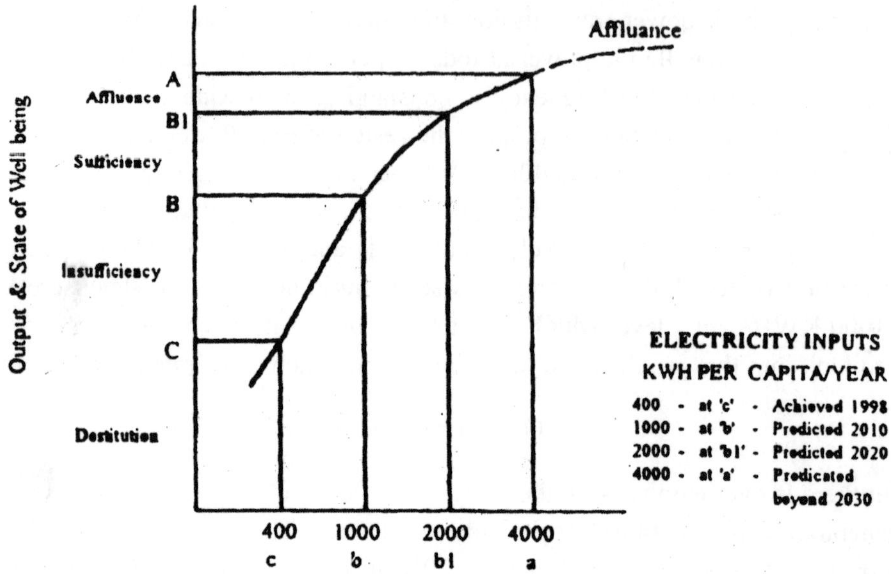

INDIAN PATTERN

Projections of Distribution of Per Capita Comsumption For Industrial, Agricultural, Domestic & Other use in the year 1998, 2010, 2020 in KWH/Capita/Year

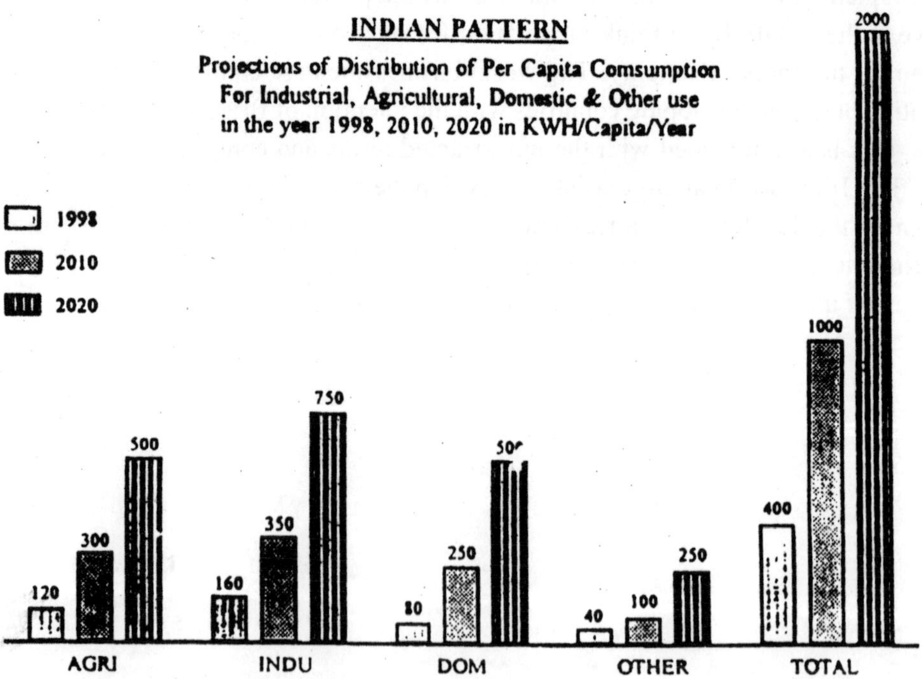

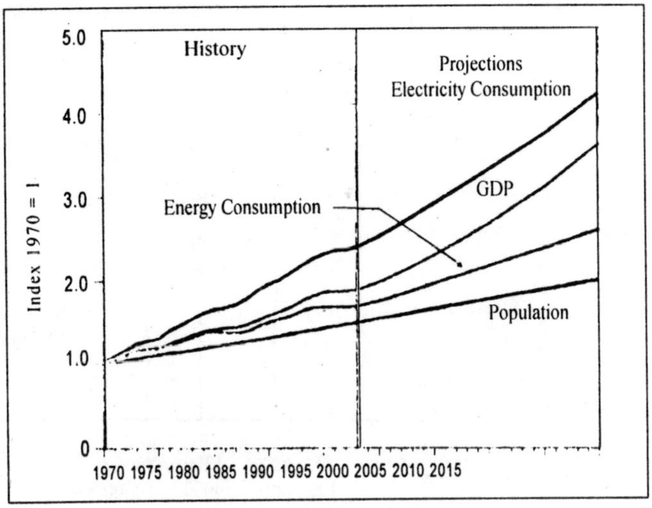

World energy, GDP, and population trends, 1970-2015

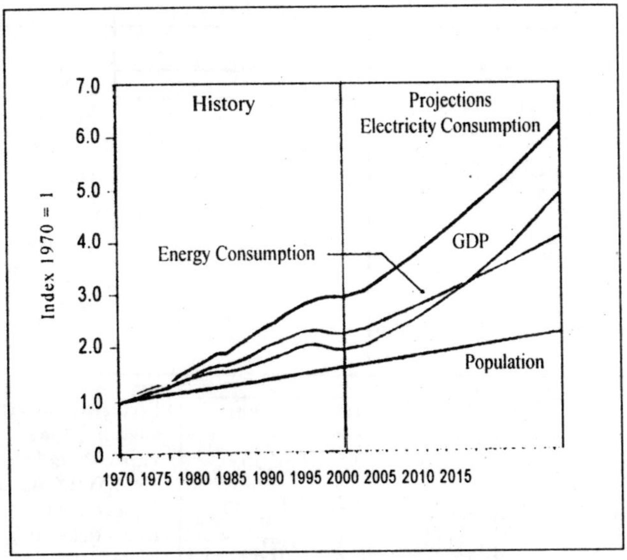

Non-OECD energy, GDP, and population

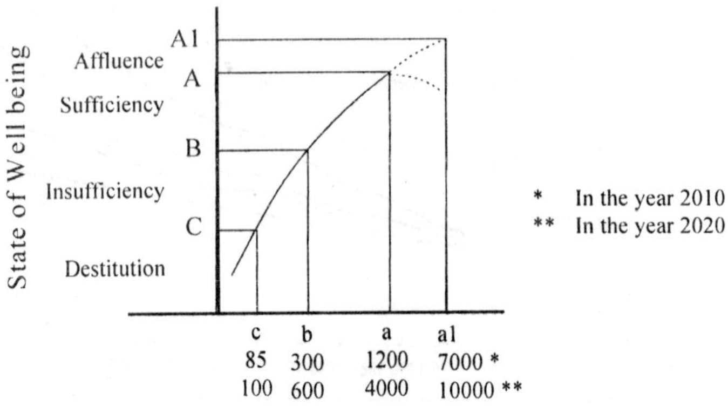

Sr. No.	*Class of Consumers*		*KWH/ Capita/Yr.*	*Total*	*Remark*
1.	*Below Poverty*	*20%*	*Not Connected*		Sr. No. 1 -contains Adivasi villages are Batti scheme. They have Financial Incapability for payment of bills of Electricity
2.	Just above Poverty	20%	85	1700	
3.	Middle Class	40%	175	7000	
4.	Upper Middle	15%	300	4500	
5.	Upper Class	4%	1200	4800	
6.	Affluent	1%	7000	7000	
min to max 1:350]	Sample of 100 Persons @ 250 KWH/Capita/Year		Year 2010	25000 KWH	

A	BPC	20	50	1,000
B	JAPC	20	100	2,000
C	MC	40	300	12,000
D	UMC	15	600	9,000
E	UC	4	4,000	16,000
F	AC	1	1,000	10,000
		100		50,000
MIN TO MAX= 1:200		SAMPLE OF 100 PERSONS		

Predicted Domestic Consumption in the year 2020 in India 500 KWH/Capita/Year In year 2020

BPC - Below Poverty Class
JAPC - Just Above Poverty Class
MC - Middle Class
UMC - Upper Middle Class
UC - Upper Class
AC - Afluent Class

FINANCIAL OF COAL PROJECT FOR POWER SECTOR IN INDIA

32

Dr. Mahendra K. Pandey[1]
Vijay K. Tiwary[2]

[1]Joint Adviser (Coal), Planning Commission, Government of India
[2]Deputy Chief Engineer, Coal India Limited

Introduction

The changing profile of industrial and economic development activities in the country, have brought in changes in the consumption pattern of coal over the last four decades. In 1960-61, the consumption of coal for generation of power was less than 10 million tonnes, and power stations consumed hardly 17% of coal produced in the country. Since then, there have been sharp increases in the consumption of coal for thermal power generation. Progressively, it increased to approximately 20% by 1970-71, to about 36% in 1980-81 to over 58% in 1990-91 and above 70% as at present.

Presently, the total installed power generating capacity in the country is about 84000 MW, generating nearly 400 BKwH of electricity annually. The power generation is predominantly through coal based stations which contribute about two-thirds of the total electricity generated., Hydel power constitutes about a quarter of the total while the rest is generated by gas and liquid fuel stations, nuclear power plants and non-conventional energy sources such as solar energy, wind energy etc. It is quite obvious that indigenous coal based power generation is going to play a very dominant role in future. The prime reason for this is the limited availability of fuel options before us. India has approximately 6% of "Proven" coal reserves available in the world against only about 0.6% reserves it has for oil and natural gas. Therefore, it hardly makes any economic sense to go for other fuel options that are primarily based on imports. The reserve position for coal in the country is given below:

Proven Reserve	:	73 billion tonnes (bt)
Indicated Reserve	:	90 bt
Inferred Reserve	:	41 bt
Total Reserve	:	204 bt

Majority of the coal reserve are suitable for power generation and have high inherent ash (approximately 40%), and are mostly located in eastern States of India.

Demand and Supply of Power Grade Coal

It is a matter of great concern that non-availability of fuel is gradually emerging as one of the biggest bottlenecks in structuring a power project. In fact, we are faced with a paradoxical situation of scarcity in the land of plenty. While on one hand we have abundant reserves of power grade coal, on the other hand, the thermal power stations (TPSs) are faced with perennial problems of shortages of coal supplies. While it is well understood that coal at the plant head has to be delivered by the railways which has its own carrying capacity limitations and bottlenecks, many a time shortages of appropriate quality coal at mine heads have been the prime reason for demand-supply mismatch. The report of the committee on Integrated Coal Policy, has underlined the demand of coal for power to be 350 mt by the terminal year of the Ninth Plan (2001-02) and to be 500 mt by the terminal year of the Tenth Plan (2006-07). This is against a total coal demand of 513 mt and 716 mt for the corresponding periods. Even in a low case situation too, having a coal demand for 325 mt during 1996-97, if one tries to calculate coal demand based on envisaged rate of growth in Gross Domestic Product (GDP) of 7% and knowing the energy GDP elasticity to be 0.9, the rate of growth for coal demand, which is 70% of the total energy demand, works out to 4.41% annually. This means the country may need 403 mt, 500 mt and 620 mt by the terminal years of the Ninth (2001-02), Tenth (2006-07) and Eleventh (2011-12) Plans respectively. The incremental production of coal from National Coal Companies (NCCs) by 2001-02 is expected to be around 60 mt and broadly it can be assumed to be adding about 12 mt per annum afterwards. This indicates a huge gap between demand and supply of domestic coal.

Beneficiation of Non Coking Coal (Power Coal)

High ash in Indian coals causes excessive wear and tear to the combustion equipment and auxiliary plants, leading to higher outages requiring increased quantum of repair and maintenance. It also means more tonnage to be transported to generate the same energy which adds to the transports costs and environmental pollution enroute. In this context, beneficiation or washing of high ash coals becomes a necessity. In case of long distance consumers, about half of the landed cost of coal at user's end goes towards payment for transport charges. Recently, Ministry of Environment and Forests (MOEF) has taken a major step although largely driven due to environmental concerns in promulgating statutory requirement of not allowing use of any coal (after year 2001 AD) having more than 34% ash for those consumers who are located more than 1000 km away from mine heads. This will make it imperative to wash/beneficiate coal which are having high ash percentages. In most other developed countries, power houses use coal having ash content in the range of 10-15% whereas Indian power houses have to use coal having ash content of about 40%. Notwithstanding Indian non coking coal having difficult washing characteristics, the technological advancements in the eighties and nineties have made it possible to economically beneficiate such coal and add to profitability of both producer and consumers.

One of the solutions to the problems in non coking coal area is perhaps provision of simple deshaling plants. With the development of 'Fluidised Bed Combustion' technology, rejects generated at such deshaling plants can now be utilised for power generation. Several such units are being set up at coal washeries in Bihar. The unit capacity of these plants is of the order of 10-12 MW. These units will not only meet the power requirement of coal washeries and mines but also can supply to the power grids and earn revenue.

Investment Requirement in Coal Sector

Coal projects are usually capital intensive and require long gestation periods. According to an estimate, the fund required for creating 10 mt per annum coal production capacity is approximately Rs. 1000 crore, with an average coal to overburden ratio of 1:1.5. Most of the Indian coal reserves which are amenable to opencast mining fall under this category. By and large, the coal sector would need well over Rs. 23,000 crore in the next five years in order to achieve even a production level of about 360 mt. The economic liberalisation policy of the government necessitates greater part of this investment requirement to be organised by own internal resources of the National Coal Companies (NCCs). The Budgetary support over the years have accordingly been minimised. It would, therefore, be imperative for the coal companies to be extremely conscious about the cost of coal production. Simultaneously, they may have to maximise capacity utilisation of machineries and improve productivity. The recent reduction in custom tariff applicable to coal imports and initiatives to allow private sector investments in coal shall obviously lead to upgradation in marketing management of the coal companies.

Capital Supply

Capital supply is capped by constraints at two levels. At national level, the constraint relates to the overall availability of capital in India. In the domestic sector, the constraint is the low savings rate, especially in the public sector. Public sector savings peaked at 4.9% of GDP in 1976/77 but declined to 0.5% in 1993/94 and were 1.7% of GDP in 1994/95. The low level of public savings give rise to the second constraint. The continuing need for the Government to borrow heavily in domestic markets, to meet its fiscal deficit "crowds out" private investment. Fiscal deficits are currently lower than the peak of 8.3% of GDP in 1990/91 though still in the range of 6% of GDP.

India has a very nominal share in the international flow of capital. Net foreign capital inflows have averaged 2.3% of GDP over 1989 to 1994. Despite this nominal share, increasing the inflow of foreign capital will be constrained by the opportunities available to investors in other countries, many of which are competitively liberalizing foreign capital controls. It is tempting to assume that the level and mix of private foreign capital available in India correctly indexes the appetite of foreign investors for Indian assets. However, information gaps and artificial capital controls are time consuming to remove and hence it is possible for proactive, investor friendly policies to vastly improve private foreign capital inflows on a sustainable basis. Clearly, though, there will be limits to the inflow of private foreign capital, which need to be quantified for forecasting purposes, our concern is with capital flows to the energy sector. This imposes the third level constraint of the intersectoral share of energy in available capital flows. This chapter does not attempt to estimate the share which will be a function of the pace of reforms in the energy sector, risk weighted returns available to investment in the sector and competing opportunities in other sectors. Since no statistically significant determinants of capital flows to the energy sector have been identified so far, this chapter stops at presenting the demand and aggregate capital supply estimates.

Foreign Capital

The stock of investable private capital in industrialized countries is assumed to grow at the same rate as World GDP. The share of such capital invested in emerging markets increases in proportion to the rising share of the GDP of developing countries in World GDP. Increasing levels of financial integration, the dilution of capital controls in developing countries and higher returns in these countries supplement this trend. A higher rate of economic growth in India as compared to the World would bring private foreign capital into India. The constraints on the supply of foreign capital is that it is capped by the share of India's GDP and India's export growth. No incremental inflows of multilateral debt have been assumed on account of the large outstanding stock of such debt. Inflows are expected to match repayments.

Domestic Capital

In the domestic market, financial liberalisation increases the assets of domestic institutional investors like Insurance companies, pension funds and mutual funds. However, the share of the private sector is limited to 25% of such assets with the rest invested in Government securities. Consistent growth increases the investable capital of domestic promoters, whose profitability increases as per GDP growth. Reform of the Public Sector industrial enterprises generates investable capital through a positive return on their assets. No direct budgetary support for capital investment from the Government to energy producers has been assumed.

How much of the estimated capital supply will be available for energy investments? The share of energy in public outlays has varied between 27% to 30% over the period 1980 to 1995. Public investments in electric power alone had a share of around 25% in total public investments. Will private investors have similar investment priorities? This chapter does not attempt an estimation of capital flows to energy sector. It is assumed that so long as energy investments are at a reasonable ratio of the GDP, risk return criterion will guide private capital to energy

Financing of Coal Projects

Contrary to general belief, investments in coal projects appear to provide the ingredients which automatically minimise risks due to following factors :

- Price stability
- Ability to get long term supply contracts
- Economic rent, value of energy in thermal coal
- Direct link to consumers
- Economies of scale—large scale mining operation
- Established transport infrastructure
- Huge reserves
- Straight forward design, estimation, feasibility
- Reliable long term demand forecasts

Even though a number of coal mine project financing, specially by Independent Power Producers (IPPs) have been facing severe problems in the recent period, it has been a fact that hardly any new coal project has been found to face financial problems. Most of the IPPs are covered through the Power Purchase Agreement (PPA). Most of the mine mouth coal development projects or captive coal mine projects are covered under the Fuel Supply Agreement (FSA) which includes price escalation formula to take care of various uncertainties. The PPA inevitably endeavours to pass through the coal price to the IPP on to its customers. Even the smallest in balance in the parameters governing coal price can be adequately covered to ensure smooth cash flow for the IPPs.

However, one problem which has been witnessed in this process is that FSAs are becoming very complex and specially where the mine is to be financed together with the power station, the FSA in one particular case was found to be over 300 pages long. Issues that had to be handled were :

— Variation in coal deliveries to IPP

— Step-in rites of the utility into mine upon IPP non-performance/default.

— Step-in rites of the IPP upon the mine's default

— Handling latent geological problems

— Mismatch of escalation formula

— Increased mine productivity sharing formula

— Coal quality bonuses/penalties

— Proportional hearing of liquidated damages for mine causing IPP start-up delay.

— Warranties on reserves-damages

— Mine equipment technology changes over time/revised mine plans.

— Environmental influences-increased cost imposed by regulatory bodies.

Investment in Coal Projects—an Attractive Proposition

Investments made in coal projects in India have been yielding attractive rate of returns after deregulation of coal prices since beginning of the nineties. Most of the national coal companies who had formulated coal projects are finding their profits going up year by year. The estimated profit of Coal India Ltd. is around Rs. 1500 crore during 1997-98. The subsidiary companies of Coal India Ltd. like Northern Coalfields Ltd., South Eastern Coalfields Ltd., Mahanadi Coalfields Ltd. and Western Coalfields Ltd. are booking very good profits quite in contrast to their financial position about decade ago.

It is quite important for a new entrant, despite all this, to go for a joint venture project with indigenously experienced coal companies with minority or majority stake to safeguard against various procedural delays that are usually said to be encountered in approval and implementation of coal projects.

Financing by Foreign Capital

Investment by foreign investors is encouraged by the Government in coal sector projects. The investment may be in the form of direct foreign investment upto a certain limit or in the form of hundred per cent capital financing through an Indian coal company. Full repatriation of profit earned through such investments is possible and liberal tax concessions measures are existing to attract investments in coal projects.

Financing by Domestic Rupee Capital

Rupee financing of coal projects is becoming extremely attractive specially due to higher dollar/rupee ratio. Coal projects are calculated to normally yield 25-30 per cent IRR in fast up coming coalfields of Talcher, Ib-Valley, Korba and many others. Rupee capital can be raised through equity financing or debt financing in the form of bonds/debentures at attractive market rates. The key, however, would be to look for such management groups which can bring up the project in the stipulated time period and avoid cost and time overrun of the project due to delay in procedural clearances, etc.

REFERENCES

Annual Report 1996-97 of the Ministry of Coal, Government of India.

Economic Survey 1995-96, 1996-97 and 1997-98, Ministry of Finance, Government of India.

Proceedings of the International Conference 'Asia Energy Vision 2020' organised by Indian Member Committee—World Energy Council, The Institution of Engineers (India) November 15-17, 1996, New Delhi.

Proceedings of the Seminar 'Coal for Power in India' organised by Indian Institute of Coal Management, Ranchi-1997.

Proceedings of the 'Pacific Ring Coal Conference'—1994, New Delhi.

Note : The views expressed in this Chapter are those of the authors and not necessarily of the organisation they belong to.

DISCUSSION

R.N. Srivastava
Chairman, Central Electricity Authority

As regards the techno-economic clearance, the provision of the Act in Section 29, 30 and 31 is based on the ceiling of investment which is dealt in Section 29. The limit prescribed now is Rs. 100 crores for the normal projects and Rs. 1,000 crores for the projects which come through the bidding route. If it exceeds the ceiling, it has to come through CEA. It will be examined under Section 30. The concurrence is finally given and the investment is tied up under Section 31. This is mandatory for all projects which supply power to consumers. In project beyond 25 MW, captive approval is done by the State government in concurrence with the CEA wherein the CEA examines the need for the project from the power supply and demand gap angle. If there is no project that fulfils the demand and supply gap as per the State government, then the CEA gives approval; and that is all. There is no thorough technique of economic appraisal for captive power plants.

P.S. Bami, Former Chairman, NTPC

There are two points for which I need clarification. All agree in the circumstance that SCRL, and NHRC are very important legislative measures. Now of course, the dilution which has taken place is in terms of discretion of the States. Assuming a particular State goes ahead with the SERC decides not to implement all the provisions then what are the sanctions or penalties to be imposed. What happens if the legal provisions are not okayed. So are we creating any legal entity without giving the teeth of imposition?

Ans : Firstly, in 1948 there is no penal provision, in 1998 Act there are provisions : fine, 3 months "imprisonment etc. Secondly, a forum has been created in the 1998 Act whereas the 1948 Act there is no forum. The SEBs, which was the supplier and the State government which was the tariff fixer, were the monopolicies. Now a forum has been created where people can go and appeal. And then an appeal has been provided that goes to the High Court. There will be wider debate of issues; people will shout "Look, you are not giving the power section 44. You cannot stop my captive power plant". The moment the captive power plant comes, there will be the situation : the Grid Power for the HT Industry would be Rs. 4 per unit and the captive power will be Rs. 2.50 per unit. Somebody will go to the forum and agitate on the provision of Section 44. It is already happening in

Orissa. A number of these issues are coming to the forum. In a democracy, unless issues come to the forum, they can never be sorted out.

A delicate issue at this stage is, "Why so IPPs not come upto the financially viable stage". CEA has expressed the opinion that everything should be controlled. That argument appear out of place at this stage of development. Development of the power sector actually calls for removal of controls. In this way, the product is instantly delivered to the market. Unless the customer pay or he is made to pay, how will the money flow back. Money has to ultimately come from the customers. So there is no point in just criticise Boards. It is time to sit together, discuss and help boards and find a solution, because the money has to come from the customer and nowhere else.